T0235678

# Advanced Courses in Mathematics
# CRM Barcelona

Centre de Recerca Matemàtica

Managing Editor:
Manuel Castellet

Giuseppe Da Prato

# Kolmogorov Equations
# for Stochastic PDEs

Birkhäuser Verlag
Basel · Boston · Berlin

Author:

Giuseppe Da Prato
Dipartimento di Matematica
Scuola Normale Superiore
Piazza dei Cavalieri 7
56126 Pisa
Italy
daprato@sns.it

2000 Mathematical Subject Classification 35K57, 35Q30, 35Q53, 37A30, 60H15

A CIP catalogue record for this book is available from the
Library of Congress, Washington D.C., USA

Bibliografische Information Der Deutschen Bibliothek
Die Deutsche Bibliothek verzeichnet diese Publikation in der Deutschen Nationalbibliografie; detaillierte
bibliografische Daten sind im Internet über <http://dnb.ddb.de> abrufbar.

ISBN 3-7643-7216-8 Birkhäuser Verlag, Basel – Boston – Berlin

This work is subject to copyright. All rights are reserved, whether the whole or part of the material is concerned, specifically the rights of translation, reprinting, re-use of illustrations, recitation, broadcasting,
reproduction on microfilms or in other ways, and storage in data banks. For any kind of use permission of the
copyright owner must be obtained.

© 2004 Birkhäuser Verlag, P.O. Box 133, CH-4010 Basel, Switzerland
Part of Springer Science+Business Media
Cover design: Micha Lotrovsky, 4106 Therwil, Switzerland
Printed on acid-free paper produced from chlorine-free pulp. TCF∞
Printed in Germany
ISBN 3-7643-7216-8

9 8 7 6 5 4 3 2 1                                                    www.birkhauser.ch

# Contents

# Preface

This book is devoted to some basic stochastic partial differential equations, in particular reaction-diffusion equations, Burgers and Navier–Stokes equations perturbed by noise.

Particular attention is paid to the corresponding Kolmogorov equations which are elliptic or parabolic equations with infinitely many variables.

The aim of the book is to present the basic elements of stochastic PDEs in a simple and self-contained way in order to cover the program of one year PhD course both in Mathematics and in Physics.

The needed prerequisites are some basic knowledge of probability, functional analysis (including fundamental properties of Gaussian measures) and partial differential equations.

This book is an expansion of a course given by the author in 1997 at the "Center de Recerca Matemàtica" in Barcelona (see [30]), which I thank for the warm hospitality.

I wish also to thank B. Goldys for reading the manuscript and making several useful comments.

This work was also supported by the research program "Analisi e controllo di equazioni di evoluzione deterministiche e stocastiche" from the Italian "Ministero della Ricerca Scientifica e Tecnologica".

Pisa, October 2004                                               Giuseppe Da Prato

# Chapter 1

# Introduction and Preliminaries

## 1.1 Introduction

We are here concerned with a stochastic differential equation in a separable Hilbert space $H$,

$$\begin{cases} dX(t,x) = (AX(t,x) + F(X(t,x)))dt + B\,dW(t), & t > 0,\ x \in H, \\ \\ X(0,x) = x, & x \in H. \end{cases} \quad (1.1)$$

Here $A \colon D(A) \subset H \to H$ is the infinitesimal generator of a strongly continuous semigroup $e^{tA}$ in $H$, $B$ is a bounded operator from another Hilbert space $U$ and $H$, $F \colon D(F) \subset H \to H$ is a nonlinear mapping and $W(t)$, $t \geq 0$, is a cylindrical Wiener process in $U$ defined in some probability space $(\Omega, \mathscr{F}, \mathbb{P})$, see Chapter 2 for a precise definition.

In applications equation (1.1) describes the evolution of an infinite dimensional dynamical system perturbed by noise (the system being considered "isolated" when $F = 0$).

In this book we shall consider several stochastic partial differential equations which can be written in the form (1.1). In each case we shall first prove existence and uniqueness of a *mild* solution. A mild solution of equation (1.1) is a mean square continuous stochastic process, adapted to $W(t)$, such that $X(t,x) \in D(F)$ for any $t \geq 0$ and

$$X(t,x) = e^{tA}x + \int_0^t e^{(t-s)A}F(X(s,x))ds + W_A(t), \quad t \geq 0, \quad (1.2)$$

where $W_A(t)$ is the *stochastic convolution* defined by

$$W_A(t) = \int_0^t e^{(t-s)A}B dW(s), \quad t \geq 0. \quad (1.3)$$

Moreover, we shall study several properties of the *transition semigroup* $P_t$ defined by [1]

$$P_t\varphi(x) = \mathbb{E}[\varphi(X(t,x))], \quad \varphi \in B_b(H), \ t \geq 0, \ x \in H, \tag{1.4}$$

as Feller and strong Feller properties and irreducibility. We recall that $P_t$ is *Feller* if $P_t\varphi$ is continuous for all $t \geq 0$ and any continuous and bounded function $\varphi$, *strong Feller* if $P_t\varphi$ is continuous for all $t > 0$ and all $\varphi \in B_b(H)$. Moreover, $P_t$ is *irreducible* if $P_t 1_I(x) > 0$ for all $x \in H$ and all open sets $I$, where $1_I$ is the characteristic function of $I$ [2].

To study asymptotic properties of the transition semigroup $P_t$ an important tool is provided by *invariant measures*. A Borel probability measure $\nu$ in $H$ is said to be *invariant* for $P_t$ if

$$\int_H P_t\varphi d\nu = \int_H \varphi d\nu \tag{1.5}$$

for all continuous and bounded functions $\varphi\colon H \to \mathbb{R}$.

If $P_t$ is irreducible, then any invariant measure $\nu$ is *full*, that is we have $\nu(B(x,r)) > 0$ for any ball $B(x,r)$ of center $x \in H$ and radius $r$. In fact from (1.5) it follows that

$$\nu(B(x,r)) = \int_H P_t 1_I(x)\nu(dx) > 0.$$

If $P_t$ is at the same time irreducible and strong Feller, then there is at most one invariant measure in view of the Doob theorem, see Theorem 1.12 [3].

We shall prove, under suitable assumptions, existence (and in some cases uniqueness) of an invariant measure $\nu$. As it is well known, this allows us to extend uniquely $P_t$ to a strongly continuous semigroup of contractions in $L^2(H,\nu)$ (still denoted $P_t$). We shall denote by $K_2$ its infinitesimal generator.

Particular attention will be paid to describing the relationship between $K_2$ and the concrete differential operator $K_0$ defined by

$$K_0\varphi(x) = \frac{1}{2} \operatorname{Tr} [CD^2\varphi(x)] + \langle Ax + F(x), D\varphi(x)\rangle, \quad \varphi \in \mathscr{E}_A(H), \tag{1.6}$$

where Tr denotes the trace, $C = BB^*$ ($B^*$ being the adjoint of $B$), and $D$ denotes the derivative with respect to $x$. Moreover, $\mathscr{E}_A(H)$ is the linear span of all real and imaginary parts of *exponential functions* $\varphi_h$,

$$\varphi_h(x) := e^{i\langle h,x\rangle}, \quad x \in H, \quad h \in D(A^*),$$

where $A^*$ is the adjoint of $A$. It is easy to see that the space $\mathscr{E}_A(H)$ is dense in $L^2(H,\nu)$. The reason for taking $h \in D(A^*)$ is that this fact is necessary in order

---

[1] $B_b(H)$ is the space of all bounded and Borel real functions in $H$.

[2] $1_I(x) = 1$ if $x \in I$, $1_I(x) = 0$ if $x \notin I$.

[3] Another powerful method to prove the uniqueness of an invariant measure is based on coupling, see [54], [74], [73], [68], [40].

that $K_0\varphi_h$ be meaningful. In fact, if $\varphi_h(x) = e^{i\langle h, x\rangle}$ we have

$$K_0\varphi_h(x) = -\left(\frac{1}{2}|C^{1/2}h|^2 + i\langle x, A^*h\rangle + i\langle F(x), h\rangle\right)\varphi_h(x), \quad x \in H.$$

So, $K_0\varphi_h$ belongs to $L^2(H, \nu)$ provided

$$x \mapsto \langle x, A^*h\rangle \text{ and } x \mapsto \langle F(x), h\rangle \in L^2(H, \nu). \tag{1.7}$$

It is not difficult, by using the Itô formula, to show that $K_2$ is an extension of $K_0$. More difficult (in some cases still an open problem) is to show that $K_2$ is the closure of $K_0$ or, equivalently, that $\mathscr{E}_A(H)$ is a core for $K_2$. When this is the case, one can prove existence and uniqueness of a *strong* solution (in the sense of Friedrichs) of the Kolmogorov equation

$$\lambda\varphi - K_0\varphi = f, \tag{1.8}$$

where $\lambda > 0$ and $f \in L^2(H, \nu)$ are given. This means that for any $\lambda > 0$ and any $f \in L^2(H, \nu)$, there exists a sequence $\{\varphi_n\} \subset \mathscr{E}_A(H)$ such that

$$\lim_{n\to\infty}\varphi_n \to \varphi, \quad \lim_{n\to\infty}(\lambda\varphi_n - K_0\varphi_n) \to f \text{ in } L^2(H, \nu).$$

This result has several important consequences. In particular the following integration by parts formula (called in French the " ídentité du carré du champs") holds,

$$\int_H K_2\varphi\,\varphi\,d\nu = -\frac{1}{2}\int_H |B^*D\varphi|^2 d\nu, \quad \varphi \in D(K_2). \tag{1.9}$$

Let us give an idea of the proof. Since we know that $\mathscr{E}_A(H)$ is a core for $K_2$, it is enough to prove (1.9) for $\varphi \in \mathscr{E}_A(H)$. In this case one can check, by a straightforward computation, the identity

$$K_0(\varphi^2) = 2K_0\varphi\,\varphi + |B^*D\varphi|^2.$$

Now, since $\nu$ is invariant, we have that $\int_H K_0(\varphi^2)d\nu = 0$, and so (1.9) follows.

Identity (1.9) implies that if $\varphi \in D(K_2)$, then $B^*D\varphi$ is well defined, so that one can study perturbation operators of the form

$$\varphi \to K\varphi + \langle G, B^*D\varphi\rangle,$$

with $G\colon H \to H$ bounded Borel. Some other interesting consequences of (1.9), such as Poincaré and log-Sobolev inequalities, will be presented later when we study specific equations.

We shall first consider the important special case when $F = 0$ (corresponding in the applications to the absence of interactions). In this case we shall write (1.1) as

$$\begin{cases} dZ(t, x) = AZ(t, x)dt + B\,dW(t), \quad t > 0, \; x \in H, \\ \\ Z(0, x) = x, \quad x \in H. \end{cases} \tag{1.10}$$

The solution $Z(t, x)$ is called the Ornstein–Uhlenbeck process. The corresponding transition semigroup will be denoted by

$$R_t\varphi(x) = \mathbb{E}[\varphi(Z(t, x))], \quad \varphi \in B_b(H).$$

If the operator $A$ is of negative type [4] it is not difficult to show that there exists a unique invariant measure $\mu$ for $R_t$. More precisely, $\mu$ is the Gaussian measure with mean 0 and covariance operator

$$Q = \int_0^\infty e^{tA} BB^* e^{tA^*} dt.$$

We shall denote by $L_2$ the infinitesimal generator of the extension of $R_t$ to $L^2(H, \mu)$ and shall prove that $L_2$ is the closure of the Kolmogorov operator

$$L_0\varphi(x) = \frac{1}{2} \text{Tr} \, [CD^2\varphi(x)] + \langle x, A^* D\varphi(x)\rangle, \quad \varphi \in \mathscr{E}_A(H). \tag{1.11}$$

However, it is useful to study the semigroup $R_t$ also in other spaces as in the space $C_b(H)$ of all uniformly continuous and bounded real functions in $H$. Here the semigroup is not strongly continuous but its infinitesimal generator $L$ can be defined, see [17], as the unique closed operator $L$ in $C_b(H)$ such that

$$(\lambda - L)^{-1} f(x) = \int_0^\infty e^{-\lambda t} R_t f(x) dt, \quad x \in H, \ \lambda > 0, \ f \in C_b(H). \tag{1.12}$$

Then we shall consider the case when $F$ is Lipschitz continuous. The results proved here will be useful to study by approximation equations with irregular coefficients.

Finally, we shall try to prove an explicit formula relating the invariant measures $\mu$ and $\nu$ of equations (1.1) and (1.10) respectively. More precisely, we shall show (under suitable assumptions), following the recent result in [35], that

$$\int_H f d\mu = \int_H f d\nu + \int_H \langle F, DL^{-1} f\rangle d\nu, \quad f \in B_b(H), \tag{1.13}$$

where $L$ is the Ornstein–Uhlenbeck generator defined by (1.12). From (1.13) it follows that $\nu$ is absolutely continuous with respect to $\mu$.

This book has an elementary character. For the sake of simplicity, we have only considered equations with additive noise and we have only studied Kolmogorov equations coming from some stochastic partial differential equations such as *reaction-diffusion* equations, *Burgers* equation and *2D-Navier–Stokes equations*.

---

[4] That is if there exists $M > 0$ and $\omega < 0$ such that $\|e^{tA}\| \leq M e^{-\omega t}$ for all $t \geq 0$.

The same method could be applied to other equations such as the wave equation [6], [26], [27], [87], [88], the Cahn–Hilliard equation [31] and the Stefan problem [7].

We mention that Kolmogorov equations can also be studied by purely analytical methods, see the monograph [51] and references therein. This method is important when one is not able to solve (1.1), see [34], [43], [44], [39].

Also in concrete equations we have not presented the more general results of the literature, which in some cases are very technical but we have used simple situations as examples.

We end this chapter by giving some preliminaries and recalling some results which will be used in what follows.

## 1.2 Preliminaries

In this book $H$ represents a separable Hilbert space (inner product $\langle \cdot, \cdot \rangle$, norm $|\cdot|$) and $L(H)$ the Banach algebra of all linear continuous operators from $H$ into $H$ endowed with the norm

$$\|T\| = \sup\{|Tx|;\ x \in H,\ |x| = 1\}, \quad T \in L(H).$$

For any $T \in L(H)$, $T^*$ is the adjoint operator of T. Moreover,

$$\Sigma(H) = \{T \in L(H) : T = T^*\}$$

and

$$L^+(H) = \{T \in \Sigma(H) : \langle Tx, x \rangle \geq 0, \quad x, y \in H\}.$$

### 1.2.1 Some functional spaces

In this section $H$ and $U$ represent separable Hilbert spaces.

- $B_b(H; U)$ is the Banach space of all bounded and Borel mappings $\varphi \colon H \to U$, endowed with the norm

$$\|\varphi\|_0 = \sup_{x \in H} |\varphi(x)|, \quad \varphi \in C_b(H).$$

- $C_b(H; U)$ is the closed subspace of $B_b(H; U)$ consisting of all uniformly continuous and bounded mappings from $H$ into $U$. If $U = \mathbb{R}$ we set $B_b(H; U) = B_b(H)$ and $C_b(H; U) = C_b(H)$.

- $C_b^1(H)$ is the space of all uniformly continuous and bounded functions $\varphi \colon H \to \mathbb{R}$ which are Fréchet differentiable on $H$ with uniformly continuous and bounded derivative $D\varphi$. We set

$$\|\varphi\|_1 = \|\varphi\|_0 + \sup_{x \in H} |D\varphi(x)|, \quad \varphi \in C_b^1(H).$$

If $\varphi \in C_b^1(H)$ and $x \in H$, we shall identify $D\varphi(x)$ with the unique element $h$ of $H$ such that

$$D\varphi(x)y = \langle h, y \rangle, \quad y \in H.$$

- $C_b^2(H)$ is the subspace of $C_b^1(H)$ of all functions $\varphi \colon H \to \mathbb{R}$ which are twice Fréchet differentiable on $H$ with uniformly continuous and bounded second derivative $D^2\varphi$. We set

$$\|\varphi\|_2 := \|\varphi\|_1 + \sup_{x \in H} \|D^2\varphi(x)\|, \quad \varphi \in C_b^2(H).$$

If $\varphi \in C_b^2(H)$ and $x \in H$, we shall identify $D^2\varphi(x)$ with the unique linear operator $T \in L(H)$ such that

$$D\varphi(x)(y, z) = \langle Ty, z \rangle, \quad y, z \in H.$$

For any $k \in \mathbb{N}$, $C_b^k(H)$ is defined in a similar way. We set finally

$$C_b^\infty(H) = \bigcap_{k=1}^{\infty} C_b^k(H).$$

- $C_b^{0,1}(H)$ is the subspace of $C_b(H)$ of all Lipschitz continuous functions. $C_b^{0,1}(H)$ is a Banach space with the norm

$$\|\varphi\|_1 := \|\varphi\|_0 + \sup \left\{ \frac{|\varphi(x) - \varphi(y)|}{|x - y|}, \ x, y \in H, \ x \neq y \right\}, \quad \varphi \in C_b^{0,1}(H).$$

- $C_b^{1,1}(H)$ is the space of all functions $\varphi \in C_b^1(H)$ such that $D\varphi$ is Lipschitz continuous. $C_b^{1,1}(H)$ is a Banach space with the norm

$$\|\varphi\|_{1,1} = \|f\|_1 + \sup \left\{ \frac{|D\varphi(x) - D\varphi(y)|}{|x - y|}, \ x, y \in H, \ x \neq y \right\}, \quad \varphi \in C_b^{1,1}(H).$$

We recall that $C_b^2(H)$ is not dense in $C_b(H)$, see [89]. The following result was proved in [75].

**Theorem 1.1.** $C_b^{1,1}(H)$ *is dense in* $C_b(H)$.

We finally consider functions having (at most) *quadratic growth*. We denote by $C_{b,2}(H)$ the space of all functions $\varphi \colon H \to \mathbb{R}$ such that the mapping

$$H \to \mathbb{R}, \ x \to \frac{\varphi(x)}{1 + |x|^2}$$

belongs to $C_b(H)$. $C_{b,2}(H)$, endowed with the norm

$$\|\varphi\|_{b,2} = \sup_{x \in H} \frac{|\varphi(x)|}{1 + |x|^2},$$

is a Banach space.

Moreover we shall denote by $C^1_{b,2}(H)$, the space of all continuously differentiable mappings $\varphi : H \to \mathbb{R}$ of $C_{b,2}(H)$ such that

$$[\varphi]_{1,2} := \sup_{x \in H} \frac{|D\varphi(x)|}{1 + |x|^2} < +\infty.$$

## 1.2.2 Exponential functions

We are here concerned with the set $\mathscr{E}(H)$ of all *exponential functions*, defined as the span of all real and imaginary parts of functions,

$$\varphi_h(x) := e^{i\langle x, h \rangle}, \ x, h \in H.$$

$\mathscr{E}(H)$ is an agebra with the usual operations.

The following approximation result of continuous functions by exponential functions will be useful in what follows. It is easy to see that the closure of $\mathscr{E}(H)$ in $C_b(H)$ does not coincide with $C_b(H)$ [5]. So we shall prove only a pointwise approximation, see [51].

**Proposition 1.2.** *For all $\varphi \in C_b(H)$, there exists a two index sequence $\{\varphi_{n_1,n_2}\} \subset \mathscr{E}(H)$ such that*

(i) $\|\varphi_{n_1,n_2}\|_0 \le \|\varphi\|_0,$

(ii) $\lim\limits_{n_1 \to \infty} \lim\limits_{n_2 \to \infty} \varphi_{n_1,n_2}(x) = \varphi(x), \quad x \in H.$

Notice that we cannot replace $\{\varphi_{n_1,n_2}\}$ with a sequence by a diagonal extraction procedure due to the pointwise character of the convergence.

*Proof.* We first assume that $H = \mathbb{R}^d$ with $d \in \mathbb{N}$. Then for any $n \in \mathbb{N}$ there exists $\psi_n \in C_b(\mathbb{R}^d)$ with the properties:

(i) $\psi_n$ is periodic with period $n$ in all its coordinates.

(ii) $\psi_n(x) = \varphi(x)$ for all $x \in [-n + 1/2, n - 1/2]^d$.

(iii) $\|\psi_n\|_0 \le \|\varphi\|_0.$

Clearly $\psi_n(x) \to \varphi(x)$ for all $x \in \mathbb{R}^d$. Moreover, by using Fourier series, we can find a sequence $\{\varphi_n\}$ in $\mathscr{E}(H)$, close to $\{\psi_n\}$ and fulfilling (i) and (ii).

Let now $H$ be infinite dimensional, $\{e_k\}$ a complete orthonormal system in $H$, and for any $m \in \mathbb{N}$ let $P_m$ be the projector on the linear space spanned by $\{e_1, \ldots, e_m\}$,

$$P_m x = \sum_{j=1}^m \langle x, e_j \rangle e_j, \quad x \in H.$$

---

[5] It is the space of all almost periodic functions in $H$.

Given $\varphi \in C_b(H)$ and $n_1 \in \mathbb{N}$, let us consider the function

$$H \to \mathbb{R}, \quad x \to \varphi(P_{n_1}x).$$

By the first part of the proof, for each $n_1 \in \mathbb{N}$, there exists a sequence $\{\varphi_{n_1,n_2}\} \subset \mathscr{E}(H)$ such that $\lim_{n_2 \to \infty} \varphi_{n_1,n_2}(x) = \varphi(P_{n_1}x)$ for all $x \in H$, and $\|\varphi_{n_1,n_2}\|_0 \leq \|\varphi\|_0$. Therefore

$$\lim_{n_1 \to \infty} \lim_{n_2 \to \infty} \varphi_{n_1,n_2}(x) = \varphi(x),$$

for all $x \in H$.                                                                                   $\square$

**Proposition 1.3.** *For all $\varphi \in C_{b,2}(H)$ there exists a two index sequence $\{\varphi_{n_1,n_2}\} \subset \mathscr{E}(H)$ such that:*

(i) $\|\varphi_{n_1,n_2}\|_{b,2} \leq \|\varphi\|_{b,2}$.

(ii) $\lim\limits_{n \to \infty} \varphi_{n_1,n_2}(x) = \varphi(x), \quad x \in H.$

*Proof.* Let first $H = \mathbb{R}^d$ and set

$$\psi(x) = \frac{\varphi(x)}{1 + |x|^2}, \quad x \in H.$$

By Proposition 1.2 there exists a sequence $\{\psi_n\} \subset \mathscr{E}(H)$ such that

(i) $\|\psi_n\|_0 \leq \|\psi\|_0 = \|\varphi\|_{b,2}$,

(ii) $\lim\limits_{n \to \infty} \psi_n(x) = \psi(x), \ x \in H.$

Setting

$$\varphi_n(x) = 1 + \sum_{i=1}^{d} (n \sin(x_i/n))^2, \quad x \in \mathbb{R}^d,$$

we have $\varphi_n \in \mathscr{E}_A(H)$, $\|\varphi_n\|_{b,2} \leq \|\varphi\|_{b,2}$ and $\lim\limits_{n \to \infty} \varphi_n(x) = \varphi(x)$, $x \in H$. If $H$ is infinite dimensional we proceed as in the second part of the proof of Proposition 1.2.                                                                                   $\square$

### 1.2.3   Gaussian measures

Let $L_1(H)$ be the Banach space of all trace class operators in $H$ endowed with the norm

$$\|T\|_1 = \mathrm{Tr}\, \sqrt{TT^*}, \quad T \in L_1(H),$$

where Tr represents the trace. We set $L_1^+(H) = L_1(H) \cap L^+(H)$. We recall that a linear operator $Q \in L^+(H)$ is of *trace class* if and only if there exists a complete orthonormal system $\{e_k\}$ in $H$ and a sequence of nonnegative numbers $\{\lambda_k\}$ such that

$$Qe_k = \lambda_k e_k, \quad k \in \mathbb{N},$$

and

$$\operatorname{Tr} Q := \sum_{k=1}^{\infty} \lambda_k < +\infty.$$

For any $a \in H$ and $Q \in L^+(H)$ we define the Gaussian probability measure $N_{a,Q}$ in $H$ by identifying $H$ with $\ell^2$ [(6)], and setting

$$N_{a,Q} = \prod_{k=1}^{\infty} N_{a_k, \lambda_k}, \quad a_k = \langle a, e_k \rangle, \ k \in \mathbb{N}.$$

In this way the measure $N_{a,Q}$ is defined on the product space $\mathbb{R}^\infty$ of all real sequences, but it is concentrated on $\ell^2$ (that is $\mu(\ell^2) = 1$) since, thanks to the monotone convergence theorem, we have

$$\int_{\mathbb{R}^\infty} |x|_{\ell^2}^2 N_{a,Q}(dx) = \sum_{k=1}^{\infty} \int_{\mathbb{R}} x_k^2 N_{a_k, \lambda_k}(dx_k) = \sum_{k=1}^{\infty} (\lambda_k + a_k^2) < +\infty.$$

If $a = 0$ we shall write $N_{a,Q} = N_Q$ for brevity. We shall always assume Ker $Q = \{0\}$ in what follows.

If $H$ is $n$-dimensional and det $Q > 0$, we have

$$N_{a,Q}(dx) = (2\pi)^{-n/2} (\det Q)^{-1/2} e^{-\frac{1}{2} \langle Q^{-1}(x-a), x-a \rangle} \, dx, \quad x \in H. \tag{1.14}$$

Let us list some useful identities. They are straightforward when $H$ is $n$-dimensional and can be easily proved in the general case letting $n \to \infty$. For $\mu = N_{a,Q}$ we have

$$\int_H |x|^2 \mu(dx) = \operatorname{Tr} Q + |a|^2, \tag{1.15}$$

$$\int_H \langle x, h \rangle \mu(dx) = \langle a, h \rangle, \quad h \in H, \tag{1.16}$$

$$\int_H \langle x - a, h \rangle \langle x - a, k \rangle \mu(dx) = \langle Qh, k \rangle, \quad h, k \in H, \tag{1.17}$$

$$\int_H e^{i\langle x, h \rangle} \mu(dx) = e^{i\langle a, h \rangle} e^{-\frac{1}{2} \langle Qh, h \rangle}, \quad h \in H. \tag{1.18}$$

The range $Q^{1/2}(H)$ of $Q^{1/2}$ is called the *Cameron–Martin* space of $N_Q$. If $H$ is infinite dimensional, $Q^{1/2}(H)$ is dense in $H$ but different from $H$ and it is important to notice that

$$N_Q(Q^{1/2}(H)) = 0. \tag{1.19}$$

---

[(6)] $\ell^2$ is the space of all sequences $\{x_k\}$ of real numbers such that $|x|_{\ell^2}^2 := \sum_{k=1}^{\infty} |x_k|^2 < +\infty$.

Let us recall the *Cameron–Martin* formula. Consider a measure $N_Q$ and the translated measure $N_{a,Q}$ with $a \in Q^{1/2}(H)$. If $H$ is finite dimensional, it follows from (1.14) that $N_{a,Q}$ and $N_Q$ are equivalent and,

$$\frac{dN_{a,Q}}{dN_Q}(x) = e^{-\frac{1}{2}|Q^{-1/2}a|^2 + \langle Q^{-1/2}a, Q^{-1/2}x \rangle}, \quad x \in H. \tag{1.20}$$

This formula does not generalize immediately in infinite dimensions. In fact in this case the term $\langle Q^{-1/2}a, Q^{-1/2}x \rangle$ is only meaningful when $x$ belongs to $Q^{1/2}(H)$ which, however, is a set having $N_Q$ measure 0 by (1.19).

To give a meaning to formula (1.20) in infinite dimensions, it is convenient to introduce the *white noise* function $W$. Consider the function

$$W \colon Q^{1/2}(H) \subset H \to L^2(H, \mu), \quad f \to W_f,$$

where

$$W_f(x) = \langle x, Q^{-1/2}f \rangle, \quad x \in H. \tag{1.21}$$

In view of (1.16) we have

$$\int_H W_f(x) W_g(x) \mu(dx) = \langle QQ^{-1/2}f, Q^{-1/2}g \rangle = \langle f, g \rangle, \quad f, g \in H.$$

Thus, $W$ is an isomorphism and, since $Q^{1/2}(H)$ is dense in $H$, it can be uniquely extended to a mapping, still denoted $W$, from the whole $H$ into $L^2(H, \mu)$.

If $f \in H$ it is usual in the literature to write

$$W_f(x) = \langle x, Q^{-1/2}f \rangle, \quad x \in H,$$

even if this is meaningful only when $f \in Q^{1/2}(H)$. We shall also follow this convention.

Now the following result can be proved by a straightforward limit procedure, see e.g. [51, Theorem 1.3.6] for details.

**Theorem 1.4.** *Let $Q \in L_1^+(H)$ and $a \in Q^{1/2}(H)$. Then the measures $N_{a,Q}$ and $N_Q$ are equivalent and*

$$\frac{dN_{a,Q}}{dN_Q}(x) = \exp\left\{-\frac{1}{2}|Q^{-1/2}a|^2 + \langle Q^{-1/2}a, Q^{-1/2}x \rangle\right\}, \quad x \in H. \tag{1.22}$$

*If $a \notin Q^{1/2}(H)$, then $N_{a,Q}$ and $N_Q$ are singular.*

Obviously the term $\langle Q^{-1/2}a, Q^{-1/2}x \rangle$ in the exponential above, should be more precisely intended as $W_{Q^{-1/2}a}(x)$.

## 1.2.4 Sobolev spaces $W^{1,2}(H,\mu)$ and $W^{2,2}(H,\mu)$

We are given a Gaussian measure $\mu = N_Q$ where $Q \in L_1^+(H)$ and $\mathrm{Ker}\, Q = \{0\}$. We denote by $\{e_n\}$ a complete orthonormal set and by $\{\lambda_n\}$ a sequence of nonnegative numbers such that

$$Qe_n = \lambda_n e_n, \quad n \in \mathbb{N}.$$

We are going to define the derivative in the sense of $L^2(H,\mu)$. We start from functions in $\mathscr{E}(H)$. Notice that $\mathscr{E}(H)$ is dense in $L^2(H,\mu)$, thanks to Proposition 1.2 and the dominated convergence theorem.

For any $\varphi \in \mathscr{E}(H)$ and $k \in \mathbb{N}$ we set

$$D_k\varphi(x) = \lim_{\varepsilon \to 0} \frac{1}{\varepsilon} [\varphi(x + \varepsilon e_k) - \varphi(x)], \quad x \in H.$$

We need an integration by parts formula.

**Lemma 1.5.** *Let $\varphi, \psi \in \mathscr{E}(H)$. Then the following identity holds.*

$$\int_H D_k\varphi\,\psi\,d\mu = -\int_H \varphi D_k\psi\,d\mu + \frac{1}{\lambda_k} \int_H x_k\varphi\psi\,d\mu. \tag{1.23}$$

*Proof.* It is enough to prove (1.23) for

$$\varphi(x) = e^{i\langle f,x\rangle}, \quad \psi(x) = e^{i\langle g,x\rangle}, \ x \in H,$$

where $f, g \in H$. Now the required identity follows from a direct computation, using (1.18). $\qquad\square$

The following corollary is an immediate consequence of (1.23).

**Corollary 1.6.** *Let $\varphi, \psi \in \mathscr{E}(H)$ and $z \in Q^{1/2}(H)$. Then the following identity holds.*

$$\int_H \langle D\varphi, z\rangle\psi d\mu = -\int_H \langle D\psi, z\rangle\varphi d\mu + \int_H \langle Q^{-1/2}z, Q^{-1/2}x\rangle\varphi\psi d\mu. \tag{1.24}$$

Obviously the first factor in the last integral has to be interpreted as

$$\langle Q^{-1/2}z, Q^{-1/2}x\rangle = W_{Q^{-1/2}z}(x).$$

**Proposition 1.7.** *The mapping*

$$D\colon \mathscr{E}(H) \subset L^2(H,\mu) \to L^2(H,\mu;H), \ \varphi \to D\varphi, \tag{1.25}$$

*is closable.*

*Proof.* Let $\{\varphi_n\} \subset \mathscr{E}(H)$ be such that

$$\varphi_n \to 0 \quad \text{in } L^2(H,\mu), \quad D\varphi_n \to F \quad \text{in } L^2(H,\mu;H).$$

We have to show that $F = 0$. Let $\psi \in \mathscr{E}(H)$ and $z \in Q^{1/2}(H)$. Then by (1.24) we have that

$$\int_H \langle D\varphi_n, z\rangle \psi d\mu = -\int_H \langle D\psi, z\rangle \varphi_n d\mu + \int_H \langle Q^{-1/2}z, Q^{-1/2}x\rangle \varphi_n \psi d\mu. \qquad (1.26)$$

Notice that, by the Hölder inequality,

$$\left| \int_H \langle Q^{-1/2}z, Q^{-1/2}x\rangle \varphi_n \psi d\mu \right|^2$$

$$\leq \|\psi\|_0^2 \int_H |\langle Q^{-1/2}z, Q^{-1/2}x\rangle|^2 d\mu \int_H \varphi_n^2 d\mu$$

$$= \|\psi\|_0^2 |Q^{-1/2}z|^2 \int_H \varphi_n^2 d\mu \to 0 \quad \text{as } n \to \infty.$$

Then, letting $n \to \infty$ in (1.26) we find that

$$\int_H \langle F(x), z\rangle \psi(x)\mu(dx) = 0.$$

This implies $F = 0$ in view of the arbitrariness of $\psi$ and $z$. $\qquad\qquad \square$

We shall still denote by $D$ the closure of the operator defined by (1.25) and by $W^{1,2}(H,\mu)$ its domain.

We now define the space $W^{2,2}(H,\mu)$. We need a result that it is a straightforward generalization of Proposition 1.7.

**Proposition 1.8.** *For any $h, k \in \mathbb{N}$ the linear operator $D_h D_k$, defined in $\mathscr{E}(H)$, is closable.*

We shall still denote by $D_h D_k$ its closure. If $\varphi$ belongs to the domain of $D_h D_k$ we say that $D_h D_k \varphi$ belongs to $L^2(H,\mu)$. Now we denote by $W^{2,2}(H,\mu)$ the linear space of all functions $\varphi \in L^2(H,\mu)$ such that $D_h D_k \varphi \in L^2(H,\mu)$ for all $h, k \in \mathbb{N}$ and

$$\sum_{h,k=1}^\infty \int_H |D_h D_k \varphi(x)|^2 \mu(dx) < +\infty.$$

Endowed with the inner product

$$\langle \varphi, \psi\rangle_{W^{2,2}(H,\mu)} = \langle \varphi, \psi\rangle_{L^2(H,\mu)} + \sum_{k=1}^\infty \int_H (D_k\varphi)(D_k\psi)d\mu$$

$$+ \sum_{h,k=1}^\infty \int_H (D_h D_k \varphi(x))(D_h D_k \psi(x))\mu(dx),$$

$W^{2,2}(H,\mu)$ is a Hilbert space.

If $\varphi \in W^{2,2}(H,\mu)$, we define $D^2\varphi$ as

$$\langle D^2\varphi(x)z, z \rangle = \sum_{h,k=1}^{\infty} D_h D_k \varphi(x) z_h z_k, \quad x, z \in H, \ \mu - \text{a.s.}$$

It is easy to see that $D^2\varphi(x)$ is an Hilbert–Schmidt operator for almost all $x \in H$ and

$$\|\varphi\|_{W^{2,2}(H,\mu)}^2 = \int_H \text{Tr}\, [(D^2\varphi(x))^2] \mu(dx).$$

## 1.2.5 Markov semigroups

We denote by $\mathscr{B}(H)$ the $\sigma$-algebra of all Borel subsets of $H$ and by $M(H)$ the space of all Borel probability measures in $H$.

A *probability kernel* $\lambda$ on $H$ is a mapping

$$[0, +\infty) \times H \to M(H), \ (t, x) \to \lambda_{t,x},$$

such that

(i) $\lambda_{t+s,x}(\Gamma) = \displaystyle\int_H \lambda_{s,y}(\Gamma)\lambda_{t,x}(dy)$ for all $t, s \geq 0, x \in H, \Gamma \in \mathscr{B}(H)$,

(ii) $\lambda_{0,x}(\Gamma) = 1_\Gamma(x)$, for all $x \in H, \Gamma \in \mathscr{B}(H)$.

Any probability kernel on $H$ defines a semigroup of linear operators $P_t$ on the space $B_b(H)$ by the formula

$$P_t\varphi(x) = \int_H \lambda_{t,x}(dy)\varphi(y), \quad t \geq 0, \ x \in H, \ \varphi \in B_b(H).$$

$P_t$ is called a *Markov semigroup*.

A Markov semigroup $P_t$ is said to be *Feller* if

$$\varphi \in C_b(H) \implies P_t\varphi \in C_b(H) \quad \text{for all } t \geq 0,$$

*strong Feller* if

$$\varphi \in B_b(H) \implies P_t\varphi \in C_b(H) \quad \text{for all } t > 0,$$

*regular* if all probabilities $\lambda_{t,x}, \ t > 0, \ x \in H$, are equivalent.

A probability measure $\mu \in M(H)$ is said to be *invariant* for the Markov semigroup $P_t$ if

$$\int_H P_t\varphi d\mu = \int_H \varphi d\mu, \quad \text{for all } t > 0 \ \varphi \in C_b(H).$$

The following theorem is due to von Neumann; for a proof see e.g. [91].

**Theorem 1.9.** *Let $P_t$ be a Markov semigroup and $\mu$ an invariant measure for $P_t$. Set*

$$M(T)\varphi(x) = \frac{1}{T}\int_0^T P_s\varphi(x)ds, \quad \varphi \in L^2(H,\mu), \ x \in H.$$

*Then there exists the limit*

$$\lim_{T\to\infty} M(T)\varphi =: M_\infty\varphi \quad \text{in } L^2(H,\mu).$$

*Moreover $M_\infty^2 = M_\infty$, $M_\infty(L^2(H,\mu)) = \Sigma$, where $\Sigma$ is the set of all stationary points of $P_t$,*

$$\Sigma = \{\varphi \in L^2(H,\mu): \ P_t\varphi = \varphi, \ t \geq 0\}$$

*and*

$$\int_H M_\infty\varphi d\mu = \int_H \varphi d\mu.$$

An invariant measure $\mu$ of $P_t$ is said to be *ergodic* if

$$\lim_{T\to+\infty} \frac{1}{T}\int_0^T P_t\varphi dt = M_\infty\varphi = \overline{\varphi}, \quad \varphi \in L^2(H,\mu),$$

where $\overline{\varphi}$ is the mean of $\varphi$,

$$\overline{\varphi} = \int_H \varphi(x)\mu(dx).$$

If the following stronger condition holds,

$$\lim_{t\to+\infty} P_t\varphi(x) = \overline{\varphi}, \quad \varphi \in L^2(H,\mu), \ x \in H, \ \mu \text{ a.e.},$$

we say that $\mu$ is *strongly mixing*.

**Proposition 1.10.** *$\mu$ is ergodic if and only if the dimension of the subspace $\Sigma$ of all stationary points of $P_t$ is 1.*

We conclude this section by stating the basic *Krylov–Bogoliubov*, *Khas'minskii* and *Doob* theorems. They play an important rôle in proving existence and uniqueness of invariant measures. For proofs see e.g. [50, Chapter 4].

Let $P_t$ be a Markov semigroup in $H$. For any $T > 0$ and any $x_0 \in H$ we denote by $\mu_{T,x_0} \in M(H)$ the mean

$$\mu_{T,x_0} = \frac{1}{T}\int_0^T \lambda_{t,x_0} dt.$$

**Theorem 1.11.** *Assume that for some $x_0 \in H$ the set $\{\mu_{T,x_0}\}_{T>0}$ is tight. Then there is an invariant measure for $P_t$.*

**Theorem 1.12.** *If $P_t$ is strong Feller and irreducible, then it possesses at most one invariant measure.*

**Theorem 1.13.** *If $P_t$ is strong Feller and irreducible, then it is strongly mixing and*

$$\lim_{t\to\infty} P_t\varphi(x) = \int_H \varphi d\mu, \quad \text{for all } \varphi \in C_b(H).$$

# Chapter 2

# Stochastic Perturbations of Linear Equations

We are given two separable Hilbert spaces $H$ and $U$ (with norms $|\cdot|$ and inner products $\langle\cdot,\cdot\rangle$), a complete orthonormal basis $\{e_k\}$ in $U$ and a sequence $\{\beta_k\}$ of mutually independent standard Brownian motions on a fixed probability space $(\Omega, \mathscr{F}, \mathbb{P})$. For any $t \geq 0$ we denote by $\mathscr{F}_t$ the $\sigma$-algebra generated by all $\beta_k(s)$ with $s \leq t$ and $k \in \mathbb{N}$.

## 2.1 Introduction

We are here concerned with the differential stochastic equation

$$
\begin{cases}
dX(t) & = & AX(t)dt + BdW(t), \quad t \geq 0, \\
\\
X(0) & = & x \in H,
\end{cases}
\tag{2.1}
$$

where $A: D(A) \subset H \to H$ and $B: U \to H$ are linear operators, and $W$ is a cylindrical Wiener process in $U$.

Formally the solution of (2.1) is given by the variation of constant formula

$$
X(t,x) = e^{tA}x + W_A(t), \quad t \geq 0,
\tag{2.2}
$$

where the process $W_A(t)$, called *stochastic convolution*, is given by

$$
W_A(t) = \int_0^t e^{(t-s)A} B dW(s), \quad t \geq 0.
$$

Always formally, the cylindrical Wiener process $W(t)$ can be defined as

$$
W(t) = \sum_{k=1}^{\infty} \beta_k(t) e_k, \quad t \geq 0.
$$

It is possible to define rigorously $W(t)$ in a suitable space larger than $H$, see e.g., [49], however we shall not use this fact, since we only need that the series

$$W_A(t) = \sum_{k=1}^{\infty} \int_0^t e^{(t-s)A} B e_k d\beta_k(s). \tag{2.3}$$

be convergent in $L^2(\Omega, \mathscr{F}, \mathbb{P})$ for all $t \geq 0$. To this purpose, we shall assume from now on that

**Hypothesis 2.1.**

(i) $A \colon D(A) \subset H \to H$ is the infinitesimal generator of a strongly continuous semigroup $e^{tA}$.

(ii) $B \in L(U; H)$.

(iii) For any $t > 0$ the linear operator $Q_t$, defined as

$$Q_t x = \int_0^t e^{sA} C e^{sA^*} x\, ds, \ \ x \in H, \ t \geq 0, \tag{2.4}$$

where $C = BB^*$ is of trace class.

By the Hille–Yosida theorem it follows that there exist $M \geq 0$ and $\omega \in \mathbb{R}$ such that

$$\|e^{tA}\| \leq M e^{\omega t}, \quad t \geq 0. \tag{2.5}$$

As we shall show in §2.2, Hypothesis 2.1–(iii) ensures that the series in (2.3) is convergent and allows us to define $W_A(t)$ as a Gaussian random variable $N_{Q_t}$. We shall also investigate some continuity properties of $W_A(t)$ which will be needed in what follows.

In Section §2.3 we shall consider the process $X(\cdot, x)$ defined by (2.2). It is called the *mild* solution of (2.1). Then we shall define the transition semigroup, called the *Ornstein–Uhlenbeck* semigroup,

$$R_t \varphi(x) = \mathbb{E}[\varphi(X(t,x))] = \int_H \varphi(y) N_{e^{tA}x, Q_t}(dy), \quad t \geq 0, \ x \in H, \ \varphi \in B_b(H).$$

We prove that $R_t$ is Feller and, when Ker $C = \{0\}$, irreducible.

Section §2.4 is devoted to proving a necessary and sufficient condition in order that $R_t$ be strong Feller, and Section §2.5 to the asymptotic behaviour of $X(t,x)$ in time and to existence and uniqueness of an invariant measure $\mu$.

In Section §2.6 we shall extend the semigroup $R_t$ to the space $L^2(H, \mu)$ and prove that a suitable subspace of $\mathscr{E}(H)$ is a core for $R_t$. Moreover we shall prove the "Carré du champs" identity and some of its consequences. Finally, §2.7 is devoted to the Poincaré and log-Sobolev inequalities and to the hypercontractivity of $R_t$ and §2.8 to some complement.

Several results of this chapter are taken from [50]. However, special attention is paid to the approximation of elements of the infinitesimal generator of $R_t$ in $C_b(H)$ by exponential functions, following [48]. These results are useful to prove essential $m$-dissipativity of several Kolmogorov operators, as we shall see in the subsequent chapters.

## 2.2 The stochastic convolution

We consider here the series (2.3) defining the stochastic convolution $W_A(t)$. The generic term of this series,

$$\int_0^t e^{(t-s)A} Be_k d\beta_k(s),$$

is a vector-valued Wiener integral. It can be defined as

$$\int_0^t e^{(t-s)A} Be_k d\beta_k(s) = \sum_{h=1}^\infty \int_0^t \langle e^{(t-s)A} Be_k, f_h \rangle d\beta_k(s) \, f_h,$$

where $\{f_h\}$ is a complete orthonormal system on $H$. It is easy to check that

$$\mathbb{E} \left| \int_0^t e^{(t-s)A} Be_k d\beta_k(s) \right|^2 = \int_0^t |e^{(t-s)A} Be_k|^2 ds.$$

**Proposition 2.2.** *Assume that Hypothesis 2.1 holds. Then for any $t \geq 0$ the series in (2.3) is convergent in $L^2(\Omega, \mathscr{F}, \mathbb{P}; H)$ to a Gaussian random variable denoted $W_A(t)$ with mean 0 and covariance operator $Q_t$, where $Q_t$ is defined by (2.4). In particular we have*

$$\mathbb{E}[|W_A(t)|^2] = \operatorname{Tr} Q_t, \quad t \geq 0.$$

*Proof.* Let $t \geq 0$, $n, p \in \mathbb{N}$. Taking into account the independence of the Brownian motions $\{\beta_i\}$, we find that

$$\mathbb{E} \left( \left| \sum_{k=n+1}^{n+p} \int_0^t e^{(t-s)A} Be_k d\beta_k(s) \right|^2 \right) = \sum_{k=n+1}^{n+p} \int_0^t |e^{(t-s)A} Be_k|^2 ds$$

$$= \sum_{k=n+1}^{n+p} \int_0^t \langle B^* e^{sA^*} e^{sA} Be_k, e_k \rangle ds.$$

Since the series $\sum_{k=1}^\infty \int_0^t \langle B^* e^{sA^*} e^{sA} Be_k, e_k \rangle ds$ is convergent to

$$\int_0^t \operatorname{Tr} [e^{sA} BB^* e^{sA^*}] ds = \operatorname{Tr} Q_t < +\infty,$$

in view of Hypothesis 2.1–(iii), it follows that the series

$$\sum_{k=1}^{\infty} \int_0^t e^{(t-s)A} Be_k d\beta_k(s)$$

is convergent in $L^2(\Omega, \mathscr{F}, \mathbb{P}; H)$ to a random variable $W_A(t)$, which, as a limit of Gaussian random variables, is Gaussian. Moreover, by a simple computation we find that

$$\mathbb{E}[\langle W_A(t), h \rangle \langle W_A(t), k \rangle] := \langle Q_t h, k \rangle, \quad h, k \in H,$$

so that the law of $W_A(t)$ is $N_{Q_t}$ as required.                                    $\square$

We study now $W_A(t)$ as a function of $t$. To this purpose, let us introduce the space

$$C_W([0, T]; L^2(\Omega, \mathscr{F}, \mathbb{P}; H)) := C_W([0, T]; H))$$

consisting of all continuous mappings $F: [0, T] \to L^2(\Omega, \mathscr{F}, \mathbb{P}; H)$ which are adapted to $W$, that is such that $F(s)$ is $\mathscr{F}_s$-measurable for any $s \in [0, T]$. $C_W([0, T]; H))$, endowed with the norm

$$\|F\|_{C_W([0,T];H))} = \left( \sup_{t \in [0,T]} \mathbb{E}\left( |F(t)|^2 \right) \right)^{1/2},$$

is a Banach space. It is called the space of all *mean square continuous adapted processes* on $[0, T]$ taking values on $H$.

**Proposition 2.3.** *Assume that Hypothesis* 2.1 *holds. Then for any $T > 0$ we have that $W_A(\cdot) \in C_W([0, T]; H)$.*

*Proof.* Write, for $t > \tau > 0$,

$$W_A(t) - W_A(\tau) = \sum_{k=1}^{\infty} \int_0^\tau [e^{(t-s)A} - e^{(\tau-s)A}] Be_k d\beta_k(s)$$

$$+ \sum_{k=1}^{\infty} \int_\tau^t e^{(t-s)A} Be_k d\beta_k(s) := I_1 + I_2.$$

Since $I_1$ and $I_2$ are independent random variables, we have that

$$\mathbb{E}|W_A(t) - W_A(\tau)|^2 = \int_0^\tau \text{Tr}\,[(e^{(t-s)A} - e^{(\tau-s)A)})C(e^{(t-s)A^*} - e^{(\tau-s)A^*})]ds$$

$$+ \int_\tau^t \text{Tr}\,[e^{sA} C e^{sA^*}]ds.$$

Consequently, $\lim_{t \to \tau} \mathbb{E}|W_A(t) - W_A(\tau)|^2 = 0$ and the conclusion follows.        $\square$

**Example 2.4 (Heat equation in an interval).** Let $H = U = L^2(0,\pi)$, $B = I$ and let $A$ be given by [1]

$$\begin{cases} D(A) = H^2(0,\pi) \cap H_0^1(0,\pi), \\ \\ Ax = D_\xi^2 x, \quad x \in D(A). \end{cases} \quad (2.6)$$

$A$ is a self-adjoint negative operator and

$$Ae_k = -k^2 e_k, \quad k \in \mathbb{N},$$

where

$$e_k(\xi) = (2/\pi)^{1/2} \sin k\xi, \quad \xi \in [0,\pi], \quad k \in \mathbb{N}.$$

Therefore in this case $Q_t$ is given by

$$Q_t x = \int_0^t e^{2sA} x \, ds = \frac{1}{2}(e^{2tA} - I)A^{-1}x, \quad x \in H.$$

Since

$$\text{Tr } Q_t = \frac{1}{2}\sum_{k=1}^\infty \frac{1 - e^{-2tk^2}}{k^2} \le \frac{1}{2}\sum_{k=1}^\infty \frac{1}{k^2} < +\infty,$$

we have that $Q_t \in L_1^+(H)$ and so, Hypothesis 2.1 is fulfilled.

**Example 2.5 (Heat equation in a square).** We consider here the heat equation in the square $\mathcal{O} = [0,\pi]^N$ with $N \in \mathbb{N}$. We choose $H = U = L^2(\mathcal{O})$, $B = I$, and set

$$\begin{cases} D(A) = H^2(\mathcal{O}) \cap H_0^1(\mathcal{O}), \\ \\ Ax = \Delta_\xi x, \quad x \in D(A), \end{cases}$$

where $\Delta_\xi$ represents the Laplace operator.

$A$ is a self-adjoint negative operator in $H$, moreover

$$Ae_k = -|k|^2 e_k, \quad k \in \mathbb{N}^N,$$

where

$$|k|^2 = k_1^2 + \cdots + k_N^2, \quad (k_1, \ldots, k_N) \in \mathbb{N}^N,$$

and

$$e_k(\xi) = (2/\pi)^{N/2} \sin k_1 \xi \cdots \sin k_N \xi, \quad \xi \in [0,\pi]^N, \; k \in \mathbb{N}^N.$$

In this case

$$\text{Tr } Q_t = \sum_{k \in \mathbb{N}^N} \frac{1}{|k|^2}\left(1 - e^{-2t|k|^2}\right) = +\infty, \quad t > 0,$$

---

[1] $H^k(0,\pi), k \in \mathbb{N}$ represent Sobolev spaces and $H_0^1(0,\pi)$ is the subspace of $H^1(0,\pi)$ of all functions vanishing at 0 and $\pi$, see e.g. [2]

for any $N > 1$.

Choose now $B = (-A)^{-\alpha/2}$, $\alpha \in (0,1)$, so that

$$Bx = \sum_{k \in \mathbb{N}^N} |k|^{-\alpha} \langle x, e_k \rangle e_k, \quad x \in H.$$

Then we have

$$\operatorname{Tr} Q_t = \sum_{k \in \mathbb{N}^N} \frac{1}{|k|^{2+2\alpha}} \left(1 - e^{-2t|k|^2}\right), \ t > 0,$$

and so, $\operatorname{Tr} Q_t < +\infty$ provided $\alpha > N/2 - 1$.

## 2.2.1 Continuity in time

We assume here that Hypothesis 2.1 is fulfilled. We know by Proposition 2.3 that $W_A(\cdot)$ is mean square continuous. In this subsection we want to show that $W_A(\cdot)(\omega)$ is continuous for $\mathbb{P}$-almost all $\omega$. In this case we say that $W_A(\cdot)$ has *continuous trajectories*. We need the following additional assumption.

**Hypothesis 2.6.** *There exists $\alpha \in (0, \frac{1}{2})$ such that*

$$\int_0^1 s^{-2\alpha} \operatorname{Tr} [e^{sA} C e^{sA^*}] ds < +\infty.$$

Note that Hypothesis 2.6 is automatically fulfilled when $C$ is of trace-class.

We shall use the *factorization method*, (see [42]) based on the elementary identity

$$\int_s^t (t - \sigma)^{\alpha-1}(\sigma - s)^{-\alpha} d\sigma = \frac{\pi}{\sin \pi\alpha}, \quad 0 \le s \le \sigma \le t, \tag{2.7}$$

where $\alpha \in (0,1)$. Using (2.7) we can write

$$W_A(t) = \frac{\sin \pi\alpha}{\pi} \int_0^t e^{(t-\sigma)A}(t - \sigma)^{\alpha-1} Y(\sigma) d\sigma, \tag{2.8}$$

where

$$Y(\sigma) = \int_0^\sigma e^{(\sigma-s)A}(\sigma - s)^{-\alpha} B dW(s), \quad \sigma \ge 0. \tag{2.9}$$

We are going to show that $Y(t)$ belongs to $L^{2m}(0, T; H)$ with probability 1. This will imply continuity of trajectories of $W_A$ by the following analytic lemma.

**Lemma 2.7.** *Let $T > 0, \alpha \in (0,1), m > 1/(2\alpha)$ and $f \in L^{2m}(0, T; H)$. Set*

$$F(t) = \int_0^t e^{(t-\sigma)A}(t - \sigma)^{\alpha-1} f(\sigma) d\sigma, \quad t \in [0, T].$$

*Then $F \in C([0, T]; H)$ and there exists a constant $C_{m,T}$ such that*

$$|F(t)| \le C_{m,T} \|f\|_{L^{2m}(0,T;H)}, \quad t \in [0, T]. \tag{2.10}$$

*Proof.* Let $M_T = \sup_{t \in [0,T]} \|e^{tA}\|$ and $t \in [0,T]$. Then by Hölder's inequality we have

$$|F(t)| \leq M_T \left( \int_0^t (t-\sigma)^{(\alpha-1)\frac{2m}{2m-1}} d\sigma \right)^{\frac{2m-1}{2m}} |f|_{L^{2m}(0,T;H)}$$

(2.11)

$$= M_T \left( \frac{2m-1}{2\alpha m - 1} \right)^{\frac{2m-1}{2m}} t^{\alpha - \frac{1}{2m}} |f|_{L^{2m}(0,T;H)},$$

that yields (2.10). It remains to show the continuity of $F$. Continuity at 0 follows from (2.11). So, it is enough to show that $F$ is continuous at any $t_0 > 0$. For $\varepsilon < \frac{t_0}{2}$ set

$$F_\varepsilon(t) = \int_0^{t-\varepsilon} e^{(t-\sigma)A}(t-\sigma)^{\alpha-1} f(\sigma) d\sigma, \quad t \in [0,T].$$

$F_\varepsilon$ is obviously continuous on $[\varepsilon, T]$. Moreover, using once again Hölder's inequality, we find that

$$|F(t) - F_\varepsilon(t)| \leq M_T \left( \frac{2m-1}{2m\alpha - 1} \right)^{\frac{2m-1}{2m}} \varepsilon^{\alpha - \frac{1}{2m}} |f|_{L^{2m}(0,T;H)}.$$

Thus $\lim_{\varepsilon \to 0} F_\varepsilon(t) = F(t)$, uniformly on $[\frac{t_0}{2}, T]$, and $F$ is continuous at $t_0$ as required. $\square$

**Exercise 2.8.** Show that Hypothesis 2.6 is fulfilled in the Example 2.4.

Now we are ready to prove the almost-sure continuity of $W_A(\cdot)$.

**Theorem 2.9.** *Assume that Hypotheses 2.1 and 2.6 hold. Let $T > 0$ and $m \in \mathbb{N}$. Then there exists a constant $C^1_{m,T} > 0$ such that*

$$\mathbb{E} \left( \sup_{t \in [0,T]} |W_A(t)|^{2m} \right) \leq C^1_{m,T}.$$

(2.12)

*Moreover $W_A(\cdot)$ is $\mathbb{P}$-almost-surely continuous on $[0,T]$.*

*Proof.* Choose $\alpha \in (0, \frac{1}{2m})$ and let $Y$ be defined by (2.9). Then, for all $\sigma \in (0,T]$, $Y(\sigma)$ is a Gaussian random variable $N_{S_\sigma}$ where

$$S_\sigma x = \int_0^\sigma s^{-2\alpha} e^{sA} C e^{sA^*} x \, ds, \quad x \in H.$$

Set Tr $(S_\sigma) = C_{\alpha,\sigma}$. Then for any $m > 1$ there exists a constant $D_{m,\alpha} > 0$ such that

$$\mathbb{E} \left( |Y(\sigma)|^{2m} \right) \leq D_{m,\alpha} \sigma^m, \quad \sigma \in [0,T].$$

This implies

$$\int_0^T \mathbb{E}\left(|Y(\sigma)|^{2m}\right) d\sigma \leq \frac{D_{m,\alpha}}{m+1}\, T^{m+1},$$

so that $Y(\cdot)(\omega) \in L^{2m}(0,T;H)$ for $\mathbb{P}$-almost all $\omega \in \Omega$. Therefore, by Lemma 2.7, $W_A(\cdot)(\omega) \in C([0,T];H)$ for $\mathbb{P}$-almost all $\omega \in \Omega$. Moreover, we have

$$\sup_{t \in [0,T]} |W_A(t)|^{2m} \leq \left(\frac{C_{M,T}}{\pi}\right)^{2m} \int_0^T |Y(\sigma)|^{2m} d\sigma.$$

Now (2.12) follows taking expectation. $\qquad\qquad\qquad\qquad\qquad\qquad\square$

### 2.2.2   Continuity in space and time

Here we assume that the Hilbert spaces $H$ and $U$ coincide with a specific space of functions $L^2(\mathcal{O})$ where $\mathcal{O}$ is a bounded subset of $\mathbb{R}^N$. We set

$$W_A(t)(\xi) = W_A(t,\xi), \quad t \geq 0, \ \xi \in \mathcal{O}.$$

We want to prove that, under Hypothesis 2.10 below, $W_A(\cdot,\cdot)(\omega) \in C([0,T] \times \mathcal{O})$ for $\mathbb{P}$-almost all $\omega \in \Omega$.

**Hypothesis 2.10.**

(i) *For any $p > 1$ the semigroup $e^{tA}$ has a unique extension to a strongly continuous semigroup in $L^p(\mathcal{O})$ which we still denote $e^{tA}$.*

(ii) *There exist $r \geq 2$ and, for any $\varepsilon \in [0,1]$, $C_\varepsilon > 0$ such that*

$$|e^{tA}x|_{W^{\varepsilon,p}(\mathcal{O})} \leq C_\varepsilon t^{-\frac{\varepsilon}{r}} |x|_{L^p(\mathcal{O})} \quad \text{for all } x \in L^p(\mathcal{O}). \tag{2.13}$$

(iii) *$A$ and $C$ are diagonal with respect to the orthonormal basis $\{e_k\}$, that is there exist sequences of positive numbers $\{\beta_k\}$ and $\{\lambda_k\}$ such that*

$$Ae_k = -\beta_k e_k, \ Ce_k = \lambda_k e_k, \quad k \in \mathbb{N}.$$

*Moreover, $\beta_k \uparrow +\infty$ as $k \to \infty$.*

(iv) *For all $k \in \mathbb{N}$, $e_k \in C(\overline{\mathcal{O}})$ and there exists $\kappa > 0$ such that*

$$|e_k(\xi)| \leq \kappa, \quad k \in \mathbb{N}, \ \xi \in \overline{\mathcal{O}}. \tag{2.14}$$

(v) *There exists $\alpha \in (0,\frac{1}{2})$ such that*

$$\sum_{k=1}^\infty \lambda_k \beta_k^{2\alpha-1} < +\infty. \tag{2.15}$$

**Example 2.11.** If $\mathcal{O} = [0, \pi]$, $A$ is as in (2.6) and $Q = I$, then Hypothesis 2.10 is fulfilled with $r = 2$ and $\alpha \in (0, 1/4)$.

More generally, let $A$ be the realization of an elliptic operator of order $2m$ with Dirichlet boundary conditions in $\mathcal{O}$. Then (i) holds, (ii) holds with $r = 2m$, see e.g. [2]. As easily seen, (iv) is fulfilled when $\mathcal{O} = [0, \pi]^N$ but does not hold in general, for instance when $\mathcal{O}$ is a ball. In this case it should be replaced by a more general condition,

$$|e_k(\xi)| \leq \kappa \beta_k^g, \quad k \in \mathbb{N}, \ \xi \in \overline{\mathcal{O}},$$

for a suitable $g > 0$. See for instance [45] and [20].

To prove continuity of $W_A(t, \xi)$ on $(t, \xi)$ we need a simple analytic lemma.

**Lemma 2.12.** *Assume that Hypothesis 2.10 holds. Let* $T > 0$, $\alpha \in (0, 1/2)$, $m > \frac{1}{\alpha}$ *and* $f \in L^{2m}([0, T] \times \mathcal{O})$. *Set*

$$F(t) = \int_0^t e^{(t-\sigma)A}(t-\sigma)^{\alpha-1} f(\sigma) d\sigma, \quad t \in [0, T].$$

*Then* $F \in C([0, T] \times \mathcal{O})$ *and there exists a constant* $C_{T,m}$ *such that*

$$\sup_{t \in [0,T], \xi \in \mathcal{O}} |F(t, \xi)|^{2m} \leq C_{T,m} |f|_{L^{2m}([0,T] \times \mathcal{O})}^{2m}. \tag{2.16}$$

*Proof.* Set $\varepsilon = \frac{1}{2} \alpha r$. Taking into account (2.13) we have that

$$
\begin{aligned}
|F(t)|_{W^{\varepsilon, 2m}(\mathcal{O})} &\leq \int_0^t (t-\sigma)^{\alpha-1} |e^{(t-\sigma)A} f(\sigma)|_{W^{\varepsilon, 2m}(\mathcal{O})} d\sigma \\
&\leq C_\varepsilon \int_0^t (t-\sigma)^{\alpha/2-1} |f(\sigma)|_{L^{2m}(\mathcal{O})} d\sigma.
\end{aligned}
$$

By using Hölder's inequality and taking into account that $\frac{m(\alpha-2)}{2m-1} > -1$, we find that

$$|F(t)|_{W^{\varepsilon, 2m}(\mathcal{O})}^{2m} \leq C_\varepsilon \left( \int_0^t (t-\sigma)^{\frac{m(\alpha-2)}{2m-1}} d\sigma \right)^{2m-1} |f|_{L^{2m}([0,T] \times \mathcal{O})}^{2m}.$$

Since $\varepsilon > \frac{1}{2m}$ we obtain (2.16) as a consequence of Sobolev's embedding theorem. $\square$

We are now ready to prove

**Theorem 2.13.** *Assume that Hypotheses 2.1 and 2.10 hold. Then* $W_A(\cdot, \cdot)$ *is continuous on* $[0, T] \times \mathcal{O}$, $\mathbb{P}$-*almost surely. Moreover, if* $m > 1/\alpha$ *we have*

$$\mathbb{E} \left( \sup_{(t, \xi) \in [0, T] \times \mathcal{O}} |W_A(t, \xi)|^{2m} \right) < +\infty.$$

*Proof.* We write $W_A(t)$ as in (2.8), where $Y$ is given by (2.9) with $B = \sqrt{C}$. Let us prove that $Y \in L^p([0,T] \times \mathcal{O})$, $p \geq 2$, $\mathbb{P}$-almost surely. First we notice that for all $\sigma \in [0,T]$, $\xi \in \mathcal{O}$, we have, setting $Y(\sigma)(\xi) = Y(\sigma,\xi)$,

$$Y(\sigma,\xi) = \sum_{k=1}^{\infty} \sqrt{\lambda_k} \int_0^\sigma e^{-\beta_k(\sigma-s)}(\sigma-s)^{-\alpha} e_k(\xi) d\beta_k(\sigma).$$

Thus, $Y(\sigma,\xi)$ is a real Gaussian random variable with mean 0 and covariance $\gamma(\sigma,\xi) = \gamma$ given by

$$\gamma = \sum_{k=1}^{\infty} \lambda_k \int_0^\sigma e^{-2\beta_k s} s^{-2\alpha} |e_k(\xi)|^2 ds.$$

Taking into account (2.14) and (2.15) we see that

$$\gamma \leq \sum_{k=1}^{\infty} \lambda_k \int_0^{+\infty} e^{-2\beta_k s} s^{-2\alpha} |e_k(\xi)|^2 ds$$

$$= \kappa^2 2^{2\alpha-1} \Gamma(1-2\alpha) \sum_{k=1}^{\infty} \lambda_k \beta_k^{2\alpha-1} < +\infty.$$

Therefore there exists $C_m > 0$ such that

$$\mathbb{E}|Y(\sigma,\xi)|^{2m} \leq C_m, \quad m > 1.$$

It follows that

$$\mathbb{E} \int_0^T \int_{\mathcal{O}} |Y(\sigma,\xi)|^{2m} d\sigma \, d\xi \leq T C_m |\mathcal{O}|,$$

where $|\mathcal{O}|$ is the Lebesgue measure of $\mathcal{O}$. So $Y \in L^{2m}([0,T] \times \mathcal{O})$ and consequently $W_A \in C([0,T] \times \mathcal{O})$, $\mathbb{P}$–a. s. Now the conclusion follows taking expectation in (2.16), with $W_A$ replacing $F$ and $Y$ replacing $f$. $\qquad\square$

**Remark 2.14.** Several other estimates for the stochastic convolutions are known. Perhaps the most general is the one included in the paper [36, Proposition 2.1].

### 2.2.3   The law of the stochastic convolution

We fix here $T > 0$ and consider the stochastic convolution as a random variable on $L^2(0,T;H)$ (or in $C([0,T] \times \mathcal{O})$).

**Proposition 2.15.** $W_A(\cdot)$ *is a Gaussian random variable on* $L^2(0,T;H)$ *with mean 0 and covariance operator* $\widetilde{Q}$ *given by*

$$\widetilde{Q}h(t) = \int_0^T g(t,s)h(s)ds, \quad h \in L^2(0,T;H), \quad t \in [0,T], \qquad (2.17)$$

*where*

$$g(t, s) = \int_0^{\min\{t,s\}} e^{(s-r)A} C e^{(t-r)A^*} dr, \ t, s \in [0, T]. \tag{2.18}$$

*Moreover, if* $Ker \ (C) = \{0\}$ *the law of* $W_A(\cdot)$ *is full.*

*Proof.* It is easy to check that $W_A(\cdot)$ is a Gaussian random variable on $\mathcal{H} := L^2(0, T; H)$ with mean 0. Let us compute its covariance operator $\widetilde{Q}$. For any $h \in \mathcal{H}$ we have

$$\langle \widetilde{Q}h, h \rangle_{\mathcal{H}} = \mathbb{E}\left[|\langle W_A(\cdot), h \rangle_{\mathcal{H}}|^2\right] = \mathbb{E}\left[\left|\int_0^T \langle W_A(t), h(t) \rangle dt\right|^2\right]$$

$$= \mathbb{E}\left[\int_0^T \int_0^T \langle W_A(t), h(t) \rangle \langle W_A(s), h(s) \rangle dt \, ds\right]. \tag{2.19}$$

On the other hand, for any $x, y \in H$ we have

$$\mathbb{E}\left[\langle W_A(t), x \rangle \langle W_A(s), y \rangle\right]$$

$$= \mathbb{E} \sum_{h,k=1}^{\infty} \int_0^t \langle e^{(t-r)A} B e_h, x \rangle d\beta_k(t) \int_0^s \langle e^{(s-r)A} B e_k, y \rangle d\beta_k(s)$$

$$= \sum_{k=1}^{\infty} \int_0^{\min\{t,s\}} \langle e^{(t-r)A} B e_k, x \rangle \langle e^{(s-r)A} B e_k, y \rangle \, dr$$

$$= \int_0^{\min\{t,s\}} \langle e^{(s-r)A} C e^{(t-r)A^*} x, y \rangle \, dr.$$

Now (2.17) follows from (2.19).

Let us prove the last statement. Let $h \in L^2(0, T; H)$ be such that $\widetilde{Q}h = 0$. Then we have

$$\int_0^t g(t, s)h(s)ds + \int_s^T g(t, s)h(s)ds = 0, \quad t \in [0, T]. \tag{2.20}$$

We assume now for simplicity that $A$ is bounded (if not, one can take the inner product of (2.20) with a vector $z \in D(A^*)$ and proceed similarly).

Differentiating (2.20) with respect to $t$ yields

$$g(t, t)h(t) + \int_0^t \left(A \int_0^s e^{(t-r)A} C e^{(s-r)A^*} dr\right) h(s)ds - g(t, t)h(t)$$

$$+ \int_0^t \left(A \int_0^t e^{(t-r)A} C e^{(s-r)A^*} dr\right) h(s)ds + \int_t^T C e^{(s-t)A^*} h(s)ds = 0.$$

Therefore

$$\int_t^T Ce^{(s-t)A^*}h(s)ds = 0.$$

Finally, differentiating with respect to $t$ and taking into account that Ker $C = \{0\}$ yields $h(t) = 0$ for all $t \in [0, T]$. This proves that Ker $\widetilde{Q} = \{0\}$.                    □

**Exercise 2.16.** Assume that Hypotheses 2.1 and 2.10, hold. Prove that $W_A(\cdot)$ is a Gaussian random variable in $C([0, T] \times \mathscr{O})$ whose covariance operator is still given by (2.17)–(2.18). Show also that if Ker $C = \{0\}$ then the law of $W_A(\cdot)$ is full.

## 2.3   The Ornstein–Uhlenbeck semigroup $R_t$

### 2.3.1   General properties

We assume here that Hypothesis 2.1 holds and consider the process $X(t, x)$ defined by (2.2). It is called the *Ornstein–Uhlenbeck process*. We recall that, in view of Proposition 2.2, $X(t, x)$ is a Gaussian random variable $N_{e^{tA}x, Q_t}$, where $Q_t$ is defined by (2.4). Let us consider the corresponding *transition semigroup* $R_t$ in $B_b(H)$,

$$R_t\varphi(x) = \mathbb{E}[\varphi(X(t, x))] = \int_H \varphi(y)N_{e^{tA}x, Q_t}(dy), \quad \varphi \in B_b(H),\ t \geq 0,\ x \in H.$$

By an obvious change of variable we see that an equivalent expression for $R_t$ is provided by

$$R_t\varphi(x) = \int_H \varphi(e^{tA}x + y)N_{Q_t}(dy), \quad \varphi \in B_b(H),\ t \geq 0,\ x \in H. \tag{2.21}$$

The space $\mathscr{E}(H)$ of exponential functions is stable for $R_t$. We have in fact by (2.21) (recall that $\varphi_h(x) = e^{i\langle x, h\rangle}$, $x, h \in H$),

$$
\begin{aligned}
R_t\varphi_h(x) &= \int_H e^{i\langle e^{tA}x + y, h\rangle}N_{Q_t}(dy) = e^{i\langle e^{tA}x, h\rangle}\int_H e^{i\langle y, h\rangle}N_{Q_t}(dy) \\
&= e^{i\langle e^{tA}x, h\rangle}e^{-\frac{1}{2}\langle Q_t h, h\rangle} = e^{-\frac{1}{2}\langle Q_t h, h\rangle}\varphi_{e^{tA^*}h}(x), \quad x \in H.
\end{aligned}
\tag{2.22}
$$

**Proposition 2.17.** *Assume that Hypothesis 2.1 is fulfilled. Then the following statements hold.*

(i) *For all $t \geq 0$, $R_t$ maps $C_b(H)$ into itself (so it is Feller), and we have*

$$\|R_t\varphi\|_0 \leq \|\varphi\|_0, \quad t \geq 0,\ \varphi \in C_b(H).$$

(ii) *For all $t, s \geq 0$ and any $\varphi \in C_b(H)$ the semigroup law holds* [2],

$$R_{t+s}\varphi = R_t R_s \varphi.$$

(iii) *If $\varphi \in C_b(H)$ and $\{\varphi_{n_1,n_2}\} \subset C_b(H)$ is a two-index sequence such that $\|\varphi_{n_1,n_2}\|_0 \leq \|\varphi\|_0$ for all $n_1, n_2 \in N$ and*

$$\lim_{n_1 \to \infty} \lim_{n_2 \to \infty} \varphi_{n_1,n_2}(x) = \varphi(x) \quad \text{for all } x \in H,$$

*we have*

$$\lim_{n_1 \to \infty} \lim_{n_2 \to \infty} R_t \varphi_{n_1,n_2}(x) = R_t \varphi(x) \quad \text{for all } x \in H, \quad t \geq 0.$$

(iv) *The mapping*

$$[0, T] \times H \to \mathbb{R}, \ (t, x) \mapsto P_t \varphi(x)$$

*is continuous for all $\varphi \in C_b(H)$.*

*Proof.* (i) Let $t > 0$ be fixed. Since $\varphi \in C_b(H)$, for any $\varepsilon > 0$ there is $\delta_\varepsilon > 0$ such that

$$x, x_1 \in H, \ Me^{\omega t}|x - x_1| \leq \delta_\varepsilon \implies |\varphi(x) - \varphi(x_1)| < \varepsilon.$$

Consequently by (2.21), if $|x - x_1| \leq \frac{1}{M} e^{-\omega t} \delta_\varepsilon$ we have $|R_t \varphi(x) - R_t \varphi(x_1)| \leq \varepsilon$. So, $R_t \varphi \in C_b(H)$.

(ii) We first consider the case when $\varphi = \varphi_h$, $h \in H$. Then by (2.22) we have

$$R_{t+s}\varphi(x) = e^{-\frac{1}{2}\langle Q_{t+s}h, h \rangle} e^{i\langle e^{(t+s)A^*}h, x \rangle}, \quad x \in H, \ t, s \geq 0$$

and

$$\begin{aligned} R_t R_s \varphi(x) &= e^{-\frac{1}{2}\langle Q_t h, h \rangle} e^{-\frac{1}{2}\langle Q_s e^{tA^*}h, e^{tA^*}h \rangle} e^{i\langle e^{(t+s)A^*}h, x \rangle} \\ &= e^{-\frac{1}{2}\langle (Q_t + e^{tA}Q_s e^{tA^*})h, h \rangle} e^{i\langle e^{(t+s)A^*}h, x \rangle}, \quad x \in H, \ t, s \geq 0. \end{aligned}$$

Since $Q_t + e^{tA}Q_s e^{tA^*} = Q_{t+s}$, it follows that $R_{t+s}\varphi = R_t R_s \varphi$. So, we have proved that

$$R_{t+s}\varphi = R_t R_s \varphi, \quad \text{for all } \varphi \in \mathscr{E}_A(H), \ t, s \geq 0.$$

Let now $\varphi \in C_b(H)$. Then, by Proposition 1.2 there exists a two-index sequence $\{\varphi_{n_1,n_2}\} \subset \mathscr{E}(H)$ such that $\lim_{n_1 \to \infty} \lim_{n_2 \to \infty} \varphi_{n_1,n_2}(x) = \varphi(x)$ for all $x \in H$ and $\|\varphi_{n_1,n_2}\|_0 \leq \|\varphi\|_0$, $n_1, n_2 \in \mathbb{N}$. Since,

$$R_{t+s}\varphi_{n_1,n_2} = R_t R_s \varphi_{n_1,n_2}, \quad n_1, n_2 \in \mathbb{N}, \ t, s \geq 0,$$

---

[2] In fact, this follows from the markovianity of $X(\cdot, x)$, but we are not using this fact here.

the conclusion of (ii) follows from the dominated convergence theorem, letting $n_2 \to \infty$ and $n_1 \to \infty$. Finally, (iii) follows again from the dominated convergence theorem.

Let us finally prove (iv). Since $C_b^1(H)$ is dense in $C_b(H)$ by Theorem 1.1, it is enough to consider the case when $\varphi \in C_b^1(H)$. Let $t_0 \in [0,T]$ and $x_0 \in H$. Then for any $t \in [0,T]$ and $x \in H$ we have

$$|R_t\varphi(x) - R_{t_0}\varphi(x_0)| \leq |R_t\varphi(x) - R_t\varphi(x_0)| + |R_t\varphi(x_0) - R_{t_0}\varphi(x_0)|$$

$$\leq \|\varphi\|_1 \|e^{tA}\| |x - x_0| + |R_t\varphi(x_0) - R_{t_0}\varphi(x_0)|.$$

Moreover

$$|R_t\varphi(x_0) - R_{t_0}\varphi(x_0)| \leq \int_H |R_{t_0}\varphi(e^{(t-t_0)A}x_0 + y) - R_{t_0}\varphi(x_0)| N_{Q_{t-t_0}}(dy)$$

$$\leq \|\varphi\|_1 |e^{(t-t_0)A}x_0 - x_0| + \|\varphi\|_1 \int_H |y| N_{Q_{t-t_0}}(dy).$$

Since, by the Hölder inequality, we have

$$\left[ \int_H |y| N_{Q_{t-t_0}}(dy) \right]^2 \leq \int_H |y|^2 N_{Q_{t-t_0}}(dy) = \mathrm{Tr}\,(Q_{t-t_0}),$$

the conclusion follows.                                                        $\square$

**Exercise 2.18.** Prove that $R_t$ is a linear bounded operator from $C_b^1(H)$ into itself for all $t \geq 0$, and that, for any $\varphi \in C_b^1(H)$, we have

$$DR_t\varphi(x) = \int_H e^{tA^*} D\varphi(e^{tA}x + y) N_{Q_t}(dy), \quad t \geq 0,\ x \in H \tag{2.23}$$

and

$$|DR_t\varphi(x)| \leq M e^{\omega t} \|\varphi\|_1, \quad t \geq 0,\ x \in H.$$

We want now to study irreducibility of $R_t$. For this we need a well-known result about irreducibility of Gaussian measures.

**Lemma 2.19.** *Let $Q \in L_1^+(H)$ such that $\mathrm{Ker}\,Q = \{0\}$ and let $a \in H$. Then we have $0 < N_{a,Q}(B(x,r)) < 1$, where $B(x,r)$ is the ball of center $x$ and radius $r$.*

*Proof.* We denote by $\{f_k\}$ a complete orthonormal system in $H$ and by $\{\lambda_k\}$ a sequence of positive numbers such that

$$Qf_k = \lambda_k f_k, \quad k \in \mathbb{N}.$$

For any $x \in H$ we set $x_k = \langle x, f_k \rangle$, $k \in \mathbb{N}$.

It is enough to consider the case $x = 0$. Setting $B_r = B(0,r)$, for any $n \in \mathbb{N}$ the following inclusion obviously holds:

$$B_r \supset \left\{ x \in H : \sum_{k=1}^{n} x_k^2 \leq \frac{r^2}{2}, \ \sum_{k=n+1}^{\infty} x_k^2 < \frac{r^2}{2} \right\}.$$

Consequently

$$N_{a,Q}(B_r) \geq N_{a,Q}\left( \sum_{k=1}^{n} x_k^2 \leq \frac{r^2}{2} \right) N_{a,Q}\left( \sum_{k=n+1}^{\infty} x_k^2 < \frac{r^2}{2} \right).$$

Now the first factor is positive, thus it is enough to show that the second is positive as well. We have in fact

$$\mu\left( \sum_{k=n+1}^{\infty} x_k^2 < \frac{r^2}{2} \right) = 1 - \mu\left( \sum_{k=n+1}^{\infty} x_k^2 \geq \frac{r^2}{2} \right)$$

$$\geq 1 - \frac{2}{r^2} \sum_{k=n+1}^{\infty} \lambda_k > 0,$$

for $n$ sufficiently large. □

**Proposition 2.20.** *Assume, besides Hypothesis 2.1, that* Ker $(B^*) = \{0\}$. *Then* $R_t$ *is irreducible.*

*Proof.* Let $t > 0$, and $x \in H$, then

$$\langle Q_t x, x \rangle = \int_0^t |B^* e^{sA^*} x|^2 ds.$$

Therefore, if $Q_t x = 0$ we have $B^* x = 0$. Consequently, since Ker $B^* = \{0\}$ we have that Ker $Q_t = \{0\}$ and the conclusion follows from Lemma 2.19. □

### 2.3.2 The infinitesimal generator of $R_t$

Let us notice that $R_t$ is not a strongly continuous semigroup on $C_b(H)$ (unless $A = 0$). In fact the limit

$$\lim_{t \to 0} R_t \varphi_h(x) = e^{-\frac{1}{2}\langle Q_t h, h \rangle} \varphi_{e^{tA^*} h}(x) = \varphi_h(x), \quad x \in H,$$

is not uniform in $x$ for any $h \neq 0$ (unless $A = 0$). However, we can give, following [94], a notion of infinitesimal generator, proceding as follows. For any $t > 0$ we set

$$\Delta_t \varphi = \frac{1}{t} \left( R_t \varphi - \varphi \right), \quad \varphi \in C_b(H).$$

Then we define the *infinitesimal generator* $L$ of $R_t$ setting

$$D(L) = \left\{ \varphi \in C_b(H) : \quad \exists f \in C_b(H), \ \lim_{t \to 0^+} \Delta_t \varphi(x) = f(x), \ \forall x \in H, \right.$$

$$\left. \text{and } \sup_{t \in (0,1]} \|\Delta_t \varphi\|_0 < +\infty \right\},$$

and

$$L\varphi(x) = \lim_{t \to 0^+} \Delta_t \varphi(x) = f(x), \quad x \in H, \ \varphi \in D(L).$$

$L$ is called the *infinitesimal generator* of $R_t$ in $C_b(H)$.

In the following we shall define $R_t$ on other spaces. In some cases, to avoid confusion, we shall write $(L, C_b(H))$ instead of $L$ and $D(L, C_b(H))$ instead $D(L)$.

Let us study some basic properties of the resolvent set $\rho(L)$ and of the resolvent $R(\lambda, L) = (\lambda - L)^{-1}$ of $L$. Results and proofs are straighforward generalizations of the classical Hille–Yosida theorem (the difference being that $R_t$ is not strongly continuous but only pointwise continuous), and so they will be rapidly sketched, for details see [94].

**Proposition 2.21.** *Assume that Hypothesis 2.1 is fulfilled. Then the following statements hold.*

(i) $(0, +\infty) \subset \rho(L)$ *and we have* [3]

$$R(\lambda, L)f(x) = \int_0^{+\infty} e^{-\lambda t} R_t f(x) dt, \quad f \in C_b(H), \ \lambda > 0, \ x \in H.$$

*Moreover,*

$$\|R(\lambda, L)f\|_0 \leq \frac{1}{\lambda} \|f\|_0, \quad \lambda > 0, \ f \in C_b(H).$$

(ii) *If $f \in C_b(H)$ and $\{f_{n_1,n_2}\} \subset C_b(H)$ is a two-index sequence such that*

$$\lim_{n_1 \to \infty} \lim_{n_2 \to \infty} f_{n_1,n_2}(x) = f(x) \quad \text{for all } x \in H$$

*and $\|f_{n_1,n_2}\|_0 \leq \|f\|_0$ for all $n_1, n_2 \in \mathbb{N}$, we have*

$$\lim_{n_1 \to \infty} \lim_{n_2 \to \infty} R(\lambda, L)f_{n_1,n_2}(x) = R(\lambda, L)f(x) \quad \text{for all } x \in H.$$

*Proof.* Let $f \in C_b(H)$. Write for any $\lambda > 0$ and any $x \in H$,

$$F(\lambda)f(x) = \int_0^{+\infty} e^{-\lambda t} R_t f(x) dt.$$

---

[3] Note the integral below is only pointwise defined in general.

It is easy to check that $F(\lambda)f \in C_b(H)$. We claim that $F(\lambda)f \in D(L)$. In fact for any $h > 0$ and $x \in H$ we have

$$R_h F(\lambda)f(x) = e^{\lambda h} \int_h^{+\infty} e^{-\lambda s} R_s f(x) ds.$$

It follows that

$$
\begin{aligned}
D_h R_h F(\lambda)f(x)|_{h=0} &= \lambda \int_0^{+\infty} e^{-\lambda s} R_s f(x) ds - f(x) \\
&= \lambda F(\lambda)f(x) - f(x),
\end{aligned}
\tag{2.24}
$$

and

$$
\begin{aligned}
|R_h F(\lambda)f(x) - F(\lambda)f(x)| &\le (e^{\lambda h} - 1) \left| \int_h^{+\infty} e^{-\lambda s} R_s f(x) ds \right| \\
&+ \left| \int_0^h e^{-\lambda s} R_s f(x) ds \right| \le \|f\|_0 \left[ \frac{e^{\lambda h} - 1}{\lambda} e^{-\lambda h} + \frac{1 - e^{-\lambda h}}{\lambda} \right] \le ch,
\end{aligned}
\tag{2.25}
$$

where $c$ is a suitable constant. By (2.24), (2.25) it follows that $F(\lambda)f \in D(L)$ and $(\lambda - L)F(\lambda)f = f$.

Let us show now that if $\varphi \in D(L)$ we have $F(\lambda)(\lambda - L)\varphi = \varphi$. This will achieve the proof. We have in fact for any $x \in H$,

$$
\begin{aligned}
F(\lambda)(\lambda - L)\varphi(x) &= \int_0^{+\infty} e^{-\lambda s} R_s(\lambda \varphi(x) - L\varphi(x)) ds \\
&= \lambda F(\lambda)(x) - \int_0^{+\infty} e^{-\lambda s} D_s R_s \varphi(x) dt.
\end{aligned}
$$

Now the conclusion (i) follows by integrating by parts, whereas (ii) and (iii) are straightforward. $\qquad \square$

**Example 2.22.** If $A \ne 0$ we have

$$D(L) \cap \mathscr{E}(H) = \{0\}.$$

In fact for any $x \in H, h \in D(A^*)$ we have

$$\lim_{t \to 0^+} \Delta_t e^{i\langle h, x \rangle} = \left[ -\frac{1}{2} \langle Ch, h \rangle + i \langle A^* h, x \rangle \right] e^{i\langle h, x \rangle},$$

which is not bounded when $A \ne 0$.

**Example 2.23.** The set

$$\mathscr{I}_A(H) = \text{linear span} \left\{ \int_0^a e^{i\langle e^{sA}x, h\rangle} ds : a > 0, \ h \in D(A^*) \right\}$$

belongs to $D(L)$, and for all $\varphi \in \mathscr{I}_A(H)$ we have

$$L\varphi(x) = \frac{1}{2} \text{Tr}\, [CD^2\varphi(x)] + \langle x, A^* D\varphi(x)\rangle, \quad x \in H.$$

In fact, setting

$$\varphi(x) = \int_0^a e^{i\langle e^{sA}x, h\rangle} ds, \quad x \in H,$$

we have

$$R_t\varphi(x) = \int_0^a e^{-\frac{1}{2}\langle Q_t e^{s^*A}h, e^{s^*A}h\rangle} e^{i\langle e^{(t+s)A}x, h\rangle} ds$$

and the conclusion follows easily.

## 2.4   The case when $R_t$ is strong Feller

Here we assume, besides Hypotheses 2.1, that

**Hypothesis 2.24.** *We have*

$$e^{tA}(H) \subset Q_t^{1/2}(H), \quad t > 0. \tag{2.26}$$

For any $t > 0$ we shall set $\Lambda_t = Q_t^{-1/2} e^{tA}$, where $Q_t^{-1/2}$ is the pseudo-inverse of $Q_t^{1/2}$. By the closed graph theorem we have that $\Lambda_t \in L(H)$ for all $t > 0$.

Hypothesis 2.24 is equivalent to the null controllability of the deterministic controlled equation in $[0, T]$ see e.g. [51],

$$y'(t) = Ay(t) + Bu(t), \quad y(0) = x, \tag{2.27}$$

where $x \in H$ and $u \in L^2(0, T; H)$. Here $y$ represents the *state* and $u$ the *control*. It is well known that (2.27) has a unique mild solution given by

$$y(t) = e^{tA}x + \int_0^t e^{(t-s)A} Bu(s)ds.$$

System (2.27) is said to be *null controllable* if for any $T > 0$ there exists $u \in L^2(0, T; H)$ such that $y(T; u) = 0$. One can show, see [98], that system (2.27) is null controllable if and only if the condition (2.26) is fulfilled. In this case, for any $x \in H$, $|\Lambda_t x|^2$ is the minimal energy for driving $x$ to 0, that is

$$|\Lambda_t x|^2 = \inf \left\{ \int_0^T |u(s)|^2 ds : u \in L^2(0, T; U), \ y(T) = 0 \right\}. \tag{2.28}$$

**Remark 2.25.** It is important to notice that if $H = U$ and $B = I$ (or if $B$ has a continuous inverse), system (2.27) is always null controllable; in fact setting $u(t) = -\frac{1}{T} e^{tA}x$ one has $y(T) = 0$. Thus in this case Hypothesis 2.24 is fulfilled. Moreover, by (2.28) it follows that

$$|\Lambda_t x|^2 \le T^{-2} \int_0^T |e^{sA}x|^2 ds, \quad t > 0, \ x \in H,$$

which yields

$$\|\Lambda_t\| \le \frac{M}{\sqrt{t}} \sup_{s \in [0,t]} e^{2\omega s}, \quad t > 0. \tag{2.29}$$

If $H$ is finite dimensional, the assumption 2.24 means that $L$ is hypoelliptic. In this case we have

$$\|\Lambda_t\| \le ct^{-(k+1)/2}, \quad t > 0,$$

for some $c > 0$ and $k \in \mathbb{N}$. $L$ is elliptic if and only if $k = 0$.

In some cases we shall need the following assumption, stronger than Hypothesis 2.24.

**Hypothesis 2.26.** *The Laplace tranform $\gamma$ of $\|\Lambda_t\|$,*

$$\gamma(\alpha) := \int_0^{+\infty} e^{-\alpha t} \|\Lambda_t\| dt < +\infty, \tag{2.30}$$

*exists for all $\alpha > \omega$.*

**Example 2.27.** Let $\{e_k\}$ be a complete orthonormal system in $H$, $A$ and $B$ linear operators defined by

$$Ae_k = -k^2 e_k, \quad Be_k = k^{-\delta} e_k, \quad k \in \mathbb{N},$$

where $\delta \in (0,1)$. Then

$$\Lambda_t e_k = \frac{\sqrt{2}\, k^{1+\delta} e^{-tk^2}}{\sqrt{1 - e^{-2tk^2}}} e_k, \quad k \in \mathbb{N}, \ t > 0,$$

and Hypotheses 2.1, 2.24 and 2.26 hold, as easily checked.

The following result implies that $R_t$ is strong Feller, see [49].

**Proposition 2.28.** *Assume that Hypotheses 2.1 and 2.24 hold. Let $\varphi \in B_b(H)$. Then for all $t > 0$ we have $R_t\varphi \in C_b^1(H)$ and* [4]

$$\langle DR_t\varphi, h \rangle = \int_H \langle \Lambda_t h, Q_t^{-1/2}y \rangle \varphi(e^{tA}x + y) N_{Q_t}(dy) \tag{2.31}$$

---

[4] One can show in fact that $R_t\varphi \in C_b^\infty(H)$, see [51]

*for all $h \in H$. Moreover*

$$|DR_t\varphi(x)| \leq \|\Lambda_t\|\|\varphi\|_0, \quad x \in H. \tag{2.32}$$

*Finally, if $\varphi \in C_b(H)$ and $\{\varphi_{n_1,n_2}\} \subset C_b(H)$ is a two-index sequence such that*

$$\lim_{n_1 \to \infty} \lim_{n_2 \to \infty} \varphi_{n_1,n_2}(x) = \varphi(x), \quad x \in H,$$

*and $\|\varphi_{n_1,n_2}\|_0 \leq \|\varphi\|_0$ for all $n_1, n_2 \in \mathbb{N}$, then we have*

$$\lim_{n_1 \to \infty} \lim_{n_2 \to \infty} DR_t\varphi_{n_1,n_2}(x) = DR_t\varphi(x), \quad x \in H, \ t \geq 0.$$

*Proof.* Let $\varphi \in B_b(H)$. Then, in view of the Cameron–Martin theorem (Theorem 1.4), and Hypothesis 2.24, we have that

$$N_{e^{tA}x,Q_t} << N_{Q_t},$$

and

$$\frac{dN_{e^{tA}x,Q_t}}{dN_{Q_t}}(y) = e^{-\frac{1}{2}|\Lambda_t x|^2 + \langle \Lambda_t x, Q_t^{-1/2}y \rangle}, \quad y \in H.$$

Consequently, for all $h \in H$ and $t > 0$ we have that $R_t\varphi$ is differentiable and

$$\langle DR_t\varphi(x), h \rangle = \int_H \langle \Lambda_t h, Q_t^{-1/2}y \rangle \varphi(e^{tA}x + y) N_{Q_t}(dy), \quad h \in H.$$

To prove (2.32), let us take the square of both sides of the identity above and use the Hölder inequality. We obtain

$$|\langle DR_t\varphi(x), h \rangle|^2 \leq \int_H |\langle \Lambda_t h, Q_t^{-1/2}y \rangle|^2 N_{Q_t}(dy) \int_H |\varphi(e^{tA}x + y)|^2 N_{Q_t}(dy)$$

$$= |\Lambda_t h|^2 \int_H |\varphi(e^{tA}x + y)|^2 N_{Q_t}(dy) \leq \|\Lambda_t\|^2 \|\varphi\|_0^2 |h|^2. \tag{2.33}$$

Now the conclusion follows by the arbitrariness of $h$. The last statement follows from the dominated convergence theorem. $\qquad\square$

**Proposition 2.29.** *Assume that Hypotheses 2.1, 2.24 and 2.26 hold. Let $f \in C_b(H)$. Then $R(\lambda, L)f \in C_b^1(H)$ and*

$$|DR(\lambda, L)f(x)| \leq \gamma(\lambda)\|f\|_0, \quad x \in H. \tag{2.34}$$

*Moreover,*

$$D(L) \subset C_b^1(H), \tag{2.35}$$

*with continuous embedding.*

*Proof.* The first statement follows taking the Laplace transform in (2.32). Let us prove (2.35). Let $\varphi \in D(L)$ and set $f = (\omega + 1)\varphi - L\varphi$, so that $\varphi = R(\omega + 1, L)f$. Then from (2.34) it follows that $\varphi \in C_b^1(H)$. $\qquad\square$

**Corollary 2.30.** *Assume that Hypotheses 2.1, 2.24 and 2.26 hold. Let $\{\varphi_{n_1,n_2}\}$ be a two-index sequence such that:*

(i) $\displaystyle\lim_{n_1 \to \infty} \lim_{n_2 \to \infty} \varphi_{n_1,n_2}(x) = \varphi(x), \quad \lim_{n_1 \to \infty} \lim_{n_2 \to \infty} L\varphi_{n_1,n_2}(x) = L\varphi(x), \quad x \in H.$

(ii) $\displaystyle\sup_{n_1,n_2 \in \mathbb{N}} \{\|\varphi_{n_1,n_2}\|_0 + \|L\varphi_{n_1,n_2}\|_0\} < +\infty.$

*Then we have*

$$\lim_{n_1 \to \infty} \lim_{n_2 \to \infty} D\varphi_{n_1,n_2}(x) = D\varphi(x), \quad x \in H,$$

*and*

$$\sup_{n_1,n_2 \in \mathbb{N}} \{\|D\varphi_{n_1,n_2}\|_0\} < +\infty.$$

*Proof.* It follows from Proposition 2.28 and 2.29. $\qquad\square$

## 2.5 Asymptotic behaviour of solutions, invariant measures

We assume here that Hypothesis 2.1 holds with $\omega < 0$ and set $\omega_1 = -\omega$. Under this assumption the linear operator

$$Q_\infty x := \int_0^{+\infty} e^{tA} C e^{tA^*} x\,dt, \quad x \in H,$$

is well defined and of trace class. We have in fact

$$Q_\infty x = \sum_{k=1}^{\infty} \int_0^1 e^{(s+k-1)A} C e^{(s+k-1)A^*} x\,ds = \sum_{k=1}^{\infty} e^{(k-1)A} Q_1 e^{(k-1)A^*} x\,ds, \quad x \in H.$$

Consequently, $\operatorname{Tr} Q_\infty \le M \sum_{k=1}^{\infty} e^{-2\omega_1(k-1)} \operatorname{Tr} Q_1 < +\infty$, as claimed.

In order to study the asymptotic behaviour of the Ornstein–Uhlenbeck process $X(t,x) = e^{tA}x + W_A(t)$, it is useful to introduce the following process $Z_A(t)$,

$$Z_A(t) = \int_0^t e^{sA} B\,dW(s), \quad t \in [0, +\infty].$$

Proceding as for the proof of Proposition 2.2 we see that the law of $Z_A(t)$ coincides with that of $W_A(t)$ for all $t \ge 0$ [5].

---

[5] Obviously, the law of $Z_A(\cdot)$ on $L^2(0,1;H)$ is not equal to that of $W_A(\cdot)$

**Lemma 2.31.** *We have*

$$\lim_{t \to +\infty} Z_A(t) = Z_A(\infty) = \int_0^{+\infty} e^{sA} B dW(s) \quad \text{in } L^2(\Omega, \mathscr{F}, \mathbb{P}; H).$$

*Proof.* We have in fact for $t, h > 0$,

$$\mathbb{E}|Z_A(t+h) - Z_A(t)|^2 = \int_t^{t+h} \text{Tr} \left[ e^{sA} C e^{sA^*} \right] ds$$

$$= \int_0^h \text{Tr} \left[ e^{(t+s)A} Q_h e^{(t+s)A^*} \right] ds = \text{Tr} \left[ e^{tA} Q_h e^{tA^*} \right] \le M e^{-2\omega_1 t} \text{Tr } Q_\infty,$$

so that $\{Z_A(t)\}_{t \ge 0}$ is Cauchy in $L^2(\Omega, \mathscr{F}, \mathbb{P}; H)$ and the conclusion follows.   □

**Proposition 2.32.** *For any $\varphi \in C_b(H)$ and any $x \in H$ we have*

$$\lim_{t \to +\infty} R_t \varphi(x) = \int_H \varphi(y) N_{Q_\infty}(dy). \tag{2.36}$$

*Proof.* In fact, if $\varphi \in C_b(H)$ we have

$$R_t \varphi(x) = \mathbb{E}[\varphi(e^{tA}x + W_A(t)] = \mathbb{E}[\varphi(e^{tA}x + Z_A(t))], \quad t > 0, \ x \in H,$$

since $W_A(t)$ and $Z_A(t)$ have the same law. Letting $t$ tend to $\infty$ we find that (2.36) holds, since the law of $Z(\infty)$ is $N_{Q_\infty}$.   □

**Proposition 2.33.** *Assume that Hypotheses 2.1 (with $\omega < 0$), 2.24 and 2.26 hold. Then for any $f \in B_b(H)$ there exist the limits*

$$\lim_{\lambda \to 0} \lambda(\lambda - L)^{-1} f(x) = \int_H f d\mu, \quad x \in H, \tag{2.37}$$

$$\lim_{\lambda \to 0} D(\lambda - L)^{-1} f(x) = \int_0^{+\infty} D R_t f(x) dt := -DL^{-1} f(x), \quad x \in H. \tag{2.38}$$

*Moreover $DL^{-1} f \in C_b(H)$.*

*Proof.* For any $f \in B_b(H)$ we have

$$\lambda(\lambda - L)^{-1} f(x) = \int_0^{+\infty} e^{-\tau} R_{\tau/\lambda} f(x) d\tau,$$

and so (2.37) follows from (2.36). Let us prove (2.38). Since $\gamma(\lambda)$ is defined in $(-\omega, +\infty)$, we have

$$\int_0^{+\infty} \|\Lambda_t\| dt < +\infty.$$

Using (2.32) this implies that $|DR_t \varphi(x)|$ is integrable in $[0, +\infty)$ and that (2.38) holds.   □

From Proposition 2.32 it follows that the measure $\mu = N_{Q_\infty}$ is *ergodic* and *strongly mixing*. Moreover, it is *invariant* for the semigroup $R_t$, that is

$$\int_H R_t\varphi(x)\mu(dx) = \int_H \varphi(x)\mu(dx), \quad \forall\, \varphi \in C_b(H). \tag{2.39}$$

We have in fact the result, see [49].

**Theorem 2.34.** *Assume here that Hypothesis 2.1 holds with* $\omega < 0$. *Then* $\mu = N_{Q_\infty}$ *is the unique invariant measure for* $R_t$.

*Proof. Existence.* Assume that $\mu = N_{Q_\infty}$. To prove that $\mu$ is invariant it is enough to prove (2.39) for $\varphi_h(x) = e^{i\langle x,h\rangle}$, $h \in H$, in view of Proposition 1.2. In this case (2.39) is equivalent to the identity

$$\hat{\mu}(e^{tA^*}h)\, e^{-\frac{1}{2}\langle Q_t h,h\rangle} = \hat{\mu}(h), \quad h \in H, \tag{2.40}$$

where $\hat{\mu}$ is the Fourier tranform of $\mu$. Now (2.40) is equivalent to

$$\langle Q_\infty e^{tA^*}h, e^{tA^*}h\rangle + \langle Q_t h, h\rangle = \langle Q_\infty h, h\rangle, \quad h \in H, t \geq 0,$$

which, in turn, is equivalent to the identity

$$e^{tA}Q_\infty e^{tA^*} + Q_t = Q_\infty, \quad t \geq 0$$

which can be easily checked.

*Uniqueness.* Let $\mu$ be an invariant measure for $R_t$. Then (2.40) holds and letting $t \to \infty$ we obtain

$$\hat{\mu}(h) = e^{-\frac{1}{2}\langle Q_\infty h,h\rangle}, \quad h \in H.$$

This implies that $\mu = N_{Q_\infty}$, by the uniqueness of the Fourier transform. $\square$

**Remark 2.35.** The necessary and sufficient condition for the existence of an invariant measure for $R_t$ is that

$$\int_0^{+\infty} |B^* e^{tA^*}x|^2 dt < +\infty \quad \text{for all } x \in H.$$

See [49].

## 2.6   The transition semigroup in $L^p(H, \mu)$

We still assume that Hypothesis 2.1 holds with $\omega < 0$, we set $\omega_1 = -\omega$ and consider the invariant measure $\mu = N_{Q_\infty}$ of $R_t$.

**Proposition 2.36.** *Assume that Hypothesis 2.1 holds with* $\omega < 0$. *Then for any* $p \geq 1$, $R_t$ *has a unique extension to a strongly continuous semigroup of contractions in* $L^p(H, \mu)$ *(which we still denote by* $R_t$).

*Proof.* Let $\varphi \in C_b(H)$. By the Hölder inequality we have that

$$|R_t\varphi(x)|^p \leq \int_H |\varphi(y)|^p N_{e^{tA}x, Q_t}(dy) = R_t(|\varphi^p|)(x), \quad t > 0, \; x \in H.$$

Integrating this identity with respect to $\mu$ over $H$, and taking into account the invariance of $\mu$, yields

$$\int_H |R_t\varphi(x)|^p \mu(dx) \leq \int_H R_t(|\varphi|^p)(x)\mu(dx) = \int_H |\varphi(x)|^p \mu(dx).$$

Since $C_b(H)$ is dense in $L^p(H, \mu)$, $R_t$ is uniquely extendible to a contraction semi-group in $L^p(H, \mu)$. The strong continuity of $R_t$ follows from the dominated convergence theorem. $\qquad\square$

For all $p \geq 1$ we shall denote by $L_p$ the infinitesimal generator of $R_t$ in $L^p(H, \mu)$ and by $D(L_p)$ its domain. Let $h \in H$, then, by the very definition of the infinitesimal generator, we see that [6]

$$\varphi_h \in D(L_p) \Longleftrightarrow h \in D(A^*).$$

If $h \in D(A^*)$ we have

$$
\begin{aligned}
L_p\varphi_h(x) &= \left[ -\frac{1}{2}\langle Ch, h\rangle + i\langle A^*h, x\rangle \right] e^{i\langle h, x\rangle}\varphi_h(x) \\
&= \frac{1}{2}\operatorname{Tr}\left[CD^2\varphi_h(x)\right] + \langle x, A^*D\varphi_h(x)\rangle, \quad x \in H.
\end{aligned}
$$

It is convenient to introduce the following subspace of $\mathscr{E}(H)$,

$$\mathscr{E}_A(H) := \text{linear span }\{\varphi_h(x) = e^{i\langle h, x\rangle} : h \in D(A^*)\}.$$

It is easy to see that $\mathscr{E}_A(H)$ is stable for $R_t$ and it is dense in $L^p(H, \mu)$ for all $p \geq 1$. Moreover, a simple generalization of Proposition 1.2 (with $\mathscr{E}_A(H)$ replacing $\mathscr{E}(H)$) holds. We have in fact the following result.

**Proposition 2.37.** *For all $\varphi \in C_b(H)$ there exists a three-index sequence $\{\varphi_{n_1, n_2, n_3}\} \subset \mathscr{E}_A(H)$ such that*

(i) $\|\varphi_{n_1, n_2, n_3}\|_0 \leq \|\varphi\|_0, \quad n_1, n_2, n_3 \in \mathbb{N},$

(ii) $\lim_{n_1 \to \infty} \lim_{n_2 \to \infty} \lim_{n_3 \to \infty} \varphi_{n_1, n_2, n_3}(x) = \varphi(x), \quad x \in H.$

---

[6] Recall that $\varphi_h(x) = e^{i\langle h, x\rangle}$, $x \in H$.

*Proof.* Let $\varphi \in C_b(H)$ and $\{\varphi_{n_1, n_2}\} \subset \mathscr{E}(H)$ be a two-index sequence fulfilling the conditions (i) and (iii) of Proposition 1.2. Set

$$\varphi_{n_1, n_2, n_3}(x) = \varphi_{n_1, n_2}(n_3(n_3 - A)^{-1}x), \quad n_1, n_2, n_3 \in \mathbb{N}.$$

Then it is easy to check that the three-index sequence $\{\varphi_{n_1, n_2, n_3}\}$ fulfills (i) and (ii). $\qquad \square$

**Theorem 2.38.** *For any $p \geq 1$, $\mathscr{E}_A(H)$ is a core for $L_p$. Moreover,*

$$L_p\varphi(x) = \frac{1}{2} \operatorname{Tr} [CD^2\varphi(x)] + \langle x, A^*D\varphi(x)\rangle, \quad x \in H, \; \varphi \in \mathscr{E}_A(H).$$

*Proof.* Since $\mathscr{E}_A(H)$ is invariant for $R_t$ and dense in $L^p(H, \mu)$, it follows that it is a core for $L_p$, see [52]. The above expression of $L_p$ follows easily computing

$$\lim_{t \to 0} \frac{1}{t} (L_p\varphi_h(x) - \varphi_h(x)),$$

for $x, h \in H$. $\qquad \square$

We will now concentrate on the case when $p = 2$. Let us first prove the so-called "Carré du champs" identity.

**Proposition 2.39.** *The operator $C^{1/2}D \colon \mathscr{E}_A(H) \subset L^2(H, \mu) \to L^2(H, \mu; H)$ is uniquely extendible to a bounded operator, denoted by $D_C$, from $D(L_2)$ into $L^2(H, \nu; H)$. Moreover the following identity holds:*

$$\int_H L_2\varphi \, \varphi \, d\mu = -\frac{1}{2} \int_H |C^{1/2}D\varphi|^2 d\mu, \quad \varphi \in D(L_2). \tag{2.41}$$

*Proof.* Let first $\varphi \in \mathscr{E}_A(H)$. Then, by a straightforward computation, we see that

$$L_2(\varphi^2) = 2\varphi L_2\varphi + |C^{1/2}D\varphi|^2.$$

Integrating this identity over $H$ with respect to $\mu$ and taking into account that

$$\int_H L_2(\varphi^2)d\mu = 0,$$

by the invariance of $\mu$, yields

$$\int_H L_2\varphi \, \varphi d\mu = -\frac{1}{2} \int_H |C^{1/2}D\varphi|^2 d\mu. \tag{2.42}$$

Now, let $\varphi \in D(L_2)$. Since $\mathscr{E}_A(H)$ is a core for $L_2$, there exists a sequence $\{\varphi_n\} \subset \mathscr{E}_A(H)$ such that

$$\varphi_n \to \varphi, \quad L_2\varphi_n \to L_2\varphi \quad \text{in } L^2(H, \nu).$$

Consequently, by (2.42) it follows that

$$\int_H |C^{1/2}D(\varphi_n - \varphi_m)|^2 d\mu \le 2\int_H |L_2(\varphi_n - \varphi_m)|\, |\varphi_n - \varphi_m|\, d\mu.$$

Therefore the sequence $\{C^{1/2}D\varphi_n\}$ is Cauchy in $L^2(H,\mu;H)$ and the conclusion follows. $\qquad\square$

Let us recall that, in view of Proposition 1.7, the linear operator

$$D: \mathscr{E}_A(H) \subset L^2(H,\mu) \to L^2(H,\mu), \quad \varphi \to D\varphi,$$

is closable in $L^2(H,\mu)$. By a similar proof, see e.g. [51], one can easily show that the operator

$$Q_\infty^{1/2}D: \mathscr{E}_A(H) \subset L^2(H,\mu) \to L^2(H,\mu), \quad \varphi \to Q_\infty^{1/2}D\varphi$$

is closable as well in $L^2(H,\mu)$. We want here to give a sufficient condition in order that $C^{1/2}D$ is closable. For this we need another assumption.

**Hypothesis 2.40.** *The operator $Q_\infty^{1/2}C^{-1/2}$ has a unique bounded extension to $H$, denoted by $K$ and $\operatorname{Ker} K = \{0\}$.*

Obviously, if $C = I$ this assumption is fulfilled.

**Proposition 2.41.** *Assume that Hypothesis 2.40 is fulfilled. Then $C^{1/2}D$ is closable.*

*Proof.* Let $\{\varphi_n\} \subset \mathscr{E}_A(H)$ such that $\varphi_n \to 0$ in $L^2(H,\mu)$ and

$$C^{1/2}D\varphi_n \to F \quad \text{in } L^2(H,\mu;H). \tag{2.43}$$

We have to show that $F = 0$. Since $C^{1/2}D\varphi_n = C^{1/2}Q_\infty^{-1/2}Q_\infty^{1/2}D\varphi_n$, we have, multiplying both sides of (2.43) by $K$,

$$Q_\infty^{1/2}D\varphi_n \to K\varphi \quad \text{in } L^2(H,\mu;H).$$

Since $Q_\infty^{1/2}D$ is closable, we obtain $KF = 0$ and so $F = 0$. $\qquad\square$

We shall denote by $D_C$ the closure of $C^{1/2}D$ and by $W_C^{1,2}(H,\mu)$ its domain.

The proof of the following proposition is similar to that of Proposition 2.39 so, it is left to the reader.

**Proposition 2.42.** *Assume that Hypothesis 2.40 is fulfilled. Then $D(L_2) \subset W_C^{1,2}(H,\mu)$ and the following identity holds:*

$$\int_H L_2\varphi\, \varphi\, d\nu = -\frac{1}{2}\int_H |D_C\varphi|^2 d\nu, \quad \varphi \in D(L_2). \tag{2.44}$$

**Proposition 2.43.** *Assume that Hypothesis 2.40 is fulfilled. Let $\varphi \in L^2(H, \nu)$ and $t \geq 0$. Set $u(t, x) = R_t\varphi(x)$. Then, for any $T > 0$, we have $u \in L^2(0, T; W_C^{1,2}(H, \mu))$ and the following identity holds:*

$$\int_H (R_t\varphi)^2 \, d\nu + \int_0^t ds \int_H |D_C R_s\varphi|^2 d\nu = \int_H \varphi^2 \, d\nu. \tag{2.45}$$

*Proof.* Let first $\varphi \in D(L_2)$. Then from the Hille–Yosida theorem we have that $R_t\varphi \in D(L_2)$ for any $t \geq 0$ and moreover

$$\frac{d}{dt} R_t\varphi = L_2 R_t\varphi.$$

Multiplying both sides of this identity by $R_t\varphi$ and integrating with respect to $x$ over $H$ we find

$$\frac{1}{2}\frac{d}{dt} \int_H (R_t\varphi)^2 d\mu = \int_H L_2 R_t\varphi \, R_t\varphi d\mu.$$

Now, taking into account (2.44) yields

$$\frac{d}{dt} \int_H (R_t\varphi)^2 d\mu = -\frac{1}{2} \int_H |D_C R_t\varphi|^2 d\mu.$$

Integrating with respect to $t$ yields (2.45) for any $\varphi \in D(L_2)$. Now the conclusion follows from the density of $D(L_2)$ in $L^2(H, \nu)$. $\square$

**Corollary 2.44.** *Assume that Hypothesis 2.40 is fulfilled. Then for any $\varphi \in L^2(H, \nu)$ we have*

$$\int_H |\varphi - \bar{\varphi}|^2 \, d\mu = \int_0^{+\infty} ds \int_H |D_C R_s\varphi|^2 d\mu, \tag{2.46}$$

*where $\bar{\varphi} = \int_H \varphi d\mu$.*

*Proof.* Letting $t \to +\infty$ in (2.45) yields

$$\lim_{t\to+\infty} \int_H (R_t\varphi)^2 \, d\mu + \int_0^{+\infty} ds \int_H |D_C R_s\varphi|^2 d\mu = \int_H \varphi^2 \, d\mu.$$

On the other hand, by (2.36) we have that

$$\lim_{t\to+\infty} \int_H (R_t\varphi)^2 \, d\mu = (\bar{\varphi})^2.$$

So, the conclusion follows. $\square$

### 2.6.1  Symmetry of $R_t$

We want to give here necessary and sufficient conditions for $R_t$ to be a symmetric operator in $L^2(H, \mu)$. We follow [22]. First we prove a lemma.

**Lemma 2.45.** $Q_\infty$ is the unique solution in $L^+(H)$ of the Lyapunov equation

$$\langle Q_\infty x, A^* y\rangle + \langle A^* x, Q_\infty y\rangle = -\langle Cx, y\rangle, \quad x, y \in D(A^*). \qquad (2.47)$$

*Proof. Existence.* Let $x, y \in D(A^*)$. Then, integrating by parts we find that

$$
\begin{aligned}
\langle Q_\infty x, A^* y\rangle &= \int_0^{+\infty} \langle e^{sA} C e^{sA^*} x, A^* y\rangle ds = \int_0^{+\infty} \langle C e^{sA^*} x, \frac{d}{ds} e^{sA^*} y\rangle ds \\
&= \langle C e^{sA^*} x, e^{sA^*} y\rangle \Big|_0^\infty - \int_0^{+\infty} \langle C e^{sA^*} A^* x, e^{sA^*} y\rangle ds \\
&= -\langle Cx, y\rangle - \langle A^* x, Q_\infty y\rangle ds
\end{aligned}
$$

so that $Q_\infty$ fulfills (2.47).

*Uniqueness.* Let $X \in L^+(H)$ be such that

$$\langle Xx, A^* y\rangle + \langle A^* x, Xy\rangle = -\langle Cx, y\rangle, \quad x, y \in D(A^*).$$

Then for any $x \in D(A^*)$ we have

$$
\begin{aligned}
\frac{d}{dt} \langle X e^{tA^*} x, e^{tA^*} x\rangle &= \langle X e^{tA^*} A^* x, e^{tA^*} x\rangle + \langle X e^{tA^*} x, e^{tA^*} A^* x\rangle \\
&= -\langle C e^{tA^*} x, e^{tA^*} x\rangle.
\end{aligned}
$$

Integrating between $0$ and $t$ yields

$$\langle X e^{tA^*} x, e^{tA^*} x\rangle = \langle Xx, x\rangle - \langle Q_t x, x\rangle, \quad x \in D(A^*).$$

Letting $t \to +\infty$ we find $\langle Xx, x\rangle = \langle Q_\infty x, x\rangle$ for all $x \in D(A^*)$, which implies $X = Q_\infty$ as required.                                                            $\square$

We can now prove a basic identity, see [15] and [62].

**Proposition 2.46.** *For all* $\varphi, \psi \in \mathscr{E}_A(H)$ *the following identity holds:*

$$\int_H L_2\varphi \, \psi d\mu = \int_H \langle Q_\infty D\psi, A^* D\varphi\rangle d\mu. \qquad (2.48)$$

*Proof.* It is enough to prove (2.48) for $\varphi = \varphi_h$ and $\psi = \varphi_k$, where $h, k \in D(A^*)$. In this case we have, by a simple computation,

$$\int_H L\varphi \, \psi d\mu = \left( \langle A^* h, Q_\infty (h - k)\rangle + \frac{1}{2} |C^{1/2} h|^2 \right) e^{-\frac{1}{2}\langle Q_\infty (h-k), h-k\rangle},$$

and

$$\int_H \langle Q_\infty D\psi, A^* D\varphi \rangle d\mu = -\langle A^* h, Q_\infty k \rangle e^{-\frac{1}{2}\langle Q_\infty (h-k), h-k \rangle}.$$

Therefore (2.53) holds since

$$2\langle A^* h, Q_\infty h \rangle + |C^{1/2} h|^2 = 0,$$

in view of the Lyapunov equation (2.47). □

**Proposition 2.47.** $L_2$ *is symmetric if and only if*

$$\langle Q_\infty x, A^* y \rangle = \langle Q_\infty y, A^* x \rangle \quad \text{for all } x, y \in D(A^*) \tag{2.49}$$

*or, equivalently, if and only if* $AQ_\infty = Q_\infty A^*$. *In this case for* $\varphi, \psi \in D(L_2)$, *we have*

$$\int_H L_2\varphi \, \psi d\mu = -\frac{1}{2} \int_H \langle C^{1/2} D\psi, C^{1/2} D\varphi \rangle d\mu. \tag{2.50}$$

*Proof.* The first statement follows from (2.48). Let us prove the second one. By (2.48) and (2.49) we have

$$\begin{aligned}
\int_H L_2\varphi \, \psi d\mu &= \frac{1}{2} \int_H \langle Q_\infty D\psi, A^* D\varphi \rangle d\mu + \frac{1}{2} \int_H \langle A^* D\varphi, Q_\infty D\psi \rangle d\mu \\
&= -\frac{1}{2} \int_H \langle C^{1/2} D\psi, C^{1/2} D\varphi \rangle d\mu,
\end{aligned}$$

in view of the Lyapunov equation. □

**Remark 2.48.** It is easy to see that condition $AQ_\infty = Q_\infty A^*$ is equivalent to

$$Ce^{tA^*} = e^{tA} C, \quad \text{for all } t \geq 0.$$

Therefore if $A = A^*$, then $L_2$ is symmetric if and only if $A$ and $C$ commute.

We are now going to study a characterization of the domain $D(L_2)$ of $L_2$. For this the following identity is useful.

**Lemma 2.49.** *For all* $\varphi \in \mathcal{E}_A(H)$ *we have*

$$L_2(|C^{1/2} D\varphi|^2|) = 2\langle CD\varphi, L\varphi \rangle - 2\langle CD\varphi, A^* D\varphi \rangle + \text{Tr}\,[(CD^2\phi)^2]. \tag{2.51}$$

*Proof.* Let $\{e_k\}$ be a complete orthonormal basis in $H$ and set $D_k\varphi = \langle D\varphi, e_k \rangle$. Then we have

$$L_2(|D_k\varphi|^2) = 2D_k\varphi \, L_2(D_k\varphi) + |C^{1/2} DD_k\varphi|^2. \tag{2.52}$$

On the other hand,

$$L_2 D_k\varphi = D_k L_2\varphi - \langle e_k, A^* D\varphi \rangle.$$

Consequently by (2.52) we find that

$$L_2(|D_k\varphi|^2) = 2D_k\varphi\, L_2(D_k\varphi) - 2D_k\varphi\langle e_k, A^*D\varphi\rangle + |C^{1/2}DD_k\varphi|^2. \qquad (2.53)$$

In an analogous way we find for $h, k \in \mathbb{N}$,

$$
\begin{aligned}
L_2(D_h\varphi D_k\varphi) &= D_k\varphi\, D_h L_2\varphi \\
&\quad + D_h\varphi\, D_k L_2\varphi - D_h\varphi\langle e_k, A^*D\varphi\rangle - D_k\varphi\langle e_h, A^*D\varphi\rangle \\
&\quad + \langle C^{1/2}DD_h\varphi, C^{1/2}DD_k\varphi\rangle.
\end{aligned}
$$
$$(2.54)$$

Now the conclusion follows from (2.53)–(2.54). □

**Proposition 2.50.** *Assume that $L_2$ is symmetric, then for any $\varphi \in \mathscr{E}_A(H)$ the following identity holds:*

$$\frac{1}{2}\int_H \mathrm{Tr}\,[(CD^2\phi)^2]d\mu - \int_H \langle CD\varphi, A^*D\varphi\rangle d\mu = 2\int_H (L\varphi)^2 d\mu. \qquad (2.55)$$

*Proof.* Let $\varphi \in \mathscr{E}_A(H)$. Then, integrating (2.51) with respect to $\mu$ over $H$ and taking into account the invariance of $\mu$ gives

$$\frac{1}{2}\int_H \mathrm{Tr}\,[(CD^2\phi)^2]d\mu - \int_H \langle CD\varphi, A^*D\varphi\rangle d\mu + \int_H \langle CD\varphi, L\varphi\rangle d\mu = 0.$$

Since $L_2$ is symmetric we have by (2.50)

$$\int_H \langle CD\varphi, L\varphi\rangle d\mu = -2\int_H (L\varphi)^2 d\mu,$$

and the conclusion follows. □

**Remark 2.51.** Identity (2.55) allows us to characterize $D(L_2)$ when $A = A^*$ and $C = I$. In fact in this case (2.55) reduces to

$$\frac{1}{2}\int_H \mathrm{Tr}\,[(CD^2\phi)^2]d\mu + \int_H |(-A)^{1/2}D\varphi|^2 d\mu = 2\int_H (L_2\varphi)^2 d\mu.$$

Consequently we have

$$D(L_2) = \left\{\varphi \in W^{2,2}(H, \mu) : \int_H |(-A)^{1/2}D\varphi|^2 d\mu < +\infty\right\}.$$

## 2.7 Poincaré and log-Sobolev inequalities

We assume here that Hypothesis 2.1 holds with $C = I, M = 1, \omega < 0$ and set $\omega_1 = -\omega$. For a more general result see [25]. We follow [53] and [51].

**Proposition 2.52.** *Assume that Hypothesis 2.1 holds with $C = I, M = 1$ and $\omega < 0$. Then, for any $\varphi \in W^{1,2}(H, \mu)$ we have*

$$\int_H |\varphi - \overline{\varphi}|^2 \, d\mu \leq \frac{1}{2\omega_1} \int_H |D\varphi|^2 d\mu, \tag{2.56}$$

*where $\overline{\varphi} = \int_H \varphi d\mu$.*

*Proof.* Let $\varphi \in C_b^1(H)$. Then by (2.23) we have

$$DR_t\varphi(x) = \int_H e^{tA^*} D\varphi(e^{tA}x + y)N_{Q_t} dy, \quad t \geq 0, \ x \in H,$$

so that

$$|DR_t\varphi(x)|^2 \leq e^{-2\omega_1 t} R_t(|D\varphi|^2)(x), \quad t \geq 0, \ x \in H.$$

Now, substituting in (2.46) and taking into account the invariance of $\mu$ yields

$$\int_H |\varphi - \overline{\varphi}|^2 \, d\mu \leq \int_0^\infty e^{-2\omega_1 t} \int_H R_t(|D\varphi|^2) d\mu$$

$$= \int_0^\infty e^{-2\omega_1 t} \int_H |D\varphi|^2 d\mu \leq \frac{1}{2\omega_1} \int_H |D\varphi|^2 d\mu. \qquad \square$$

**Proposition 2.53.** *Under the assumptions of Proposition 2.52 we have*

$$\sigma(L_2) \subset \{\lambda \in \mathbb{C} : \ \operatorname{Re} \lambda \leq \omega_1\},$$

*where $\sigma(L_2)$ is the spectrum of $L_2$. Moreover,*

$$\int_H |R_t\varphi - \overline{\varphi}|^2 \leq e^{-2\omega_1 t} \int_H \varphi^2 d\mu, \quad t \geq 0.$$

*Proof.* Define

$$L_0^2(H, \mu) := \{\varphi \in L_0^2(H, \mu) : \overline{\varphi} = 0\}.$$

Clearly, $L_0^2(H, \mu)$ is stable for $R_t$. Moreover, if $\varphi \in L_0^2(H, \mu)$ we have, taking into account (2.56),

$$\int_H L_2\varphi \, \varphi d\mu = -\frac{1}{2} \int_H |D_C\varphi|^2 d\mu \leq \frac{\omega_1}{2M_1} \int_H \varphi^2 d\mu.$$

Now the conclusion follows from the Hille–Yosida theorem. $\qquad \square$

We are going to prove again following [53] and [51], the log-Sobolev inequality. For this we need a lemma.

**Lemma 2.54.** *For any* $g \in C^2(\mathbb{R})$ *and any* $\varphi \in \mathscr{E}_A(H)$, *we have*

$$L_2(g(\varphi)) = g'(\varphi)L_2\varphi + \frac{1}{2}g''(\varphi)|D\varphi|^2, \tag{2.57}$$

*and*

$$\int_H (L_2\varphi)g'(\varphi)d\mu = -\frac{1}{2}\int_H g''(\varphi)|D\varphi|^2 d\mu. \tag{2.58}$$

*Proof.* Let $\varphi \in \mathscr{E}_A(H)$. Since

$$Dg(\varphi) = g'(\varphi)D\varphi, \quad D^2g(\varphi) = g''(\varphi)D\varphi \otimes D\varphi + g'(\varphi)D^2\varphi,$$

(2.57) follows from a simple computation. Finally integrating (2.57) yields (2.58) since

$$\int_H L_2(g(\varphi))d\mu = 0,$$

by the invariance of $\mu$. $\qquad\qquad\qquad\qquad\qquad\qquad\qquad\qquad\qquad\qquad\qquad\square$

We are now ready to prove the *log-Sobolev inequality*.

**Theorem 2.55.** *Under the assumptions of Proposition* 2.53, *for all* $\varphi \in W^{1,2}(H, \mu)$ *we have*

$$\int_H \varphi^2 \log(\varphi^2)d\mu \leq \frac{1}{\omega_1}\int_H |D\varphi|^2 d\mu + \|\varphi\|_{L^2(H,\mu)}^2 \log(\|\varphi\|_{L^2(H,\mu)}^2). \tag{2.59}$$

*Proof.* It is enough to prove the result when $\varphi \in \mathscr{E}_A(H)$ is such that $\varphi(x) \geq \varepsilon > 0$, $x \in H$. In this case we have

$$\frac{d}{dt}\int_H (R_t(\varphi^2))\log(R_t(\varphi^2))d\mu = \int_H L_2R_t(\varphi^2)\log(R_t(\varphi^2))d\mu + \int_H L_2R_t(\varphi^2)d\mu.$$

Now the second term in the right-hand side vanishes, due to the invariance of $\mu$. For the first term we use (2.58) with $g'(\xi) = \log \xi$ and obtain

$$\frac{d}{dt}\int_H R_t(\varphi^2)\log(R_t(\varphi^2))d\mu = -\frac{1}{2}\int_H \frac{1}{R_t(\varphi^2)}|DR_t(\varphi^2)|^2 d\mu. \tag{2.60}$$

On the other hand, for any $x, h \in H, t > 0$ we have

$$\langle DR_t(\varphi^2)(x), h\rangle = 2\int_H \varphi(e^{tA}x + y)\langle D\varphi(e^{tA}x + y), e^{tA}h\rangle N_{Q_t}(dy).$$

Now, using the Hölder inequality, we obtain

$$|\langle DR_t(\varphi^2)(x), h\rangle|^2 \leq 4e^{-2t\omega_1} \int_H \varphi^2(e^{tA}x + y) N_{Q_t}(dy)$$

$$\times \int_H |D\varphi(e^{tA}x + y)|^2 N_{Q_t}(dy) |h|^2,$$

which yields

$$|DR_t(\varphi^2)|^2 \leq 4e^{-2t\omega_1} R_t(\varphi^2) R_t(|D\varphi|^2).$$

Substituting in (2.60) yields

$$\frac{d}{dt} \int_H R_t(\varphi^2) \log(R_t(\varphi^2)) d\mu \geq -2e^{-2t\omega_1} \int_H R_t(|D\varphi|^2) d\mu$$

$$= -2M_1 e^{-2t\omega_1} \int_H |D\varphi|^2 d\mu,$$

due to the invariance of $\mu$. Integrating in $t$ yields

$$\int_H R_t(\varphi^2) \log(R_t(\varphi^2)) d\mu - \int_H \varphi^2 \log(\varphi^2) d\mu \geq \frac{1}{\omega_1}(1 - e^{-2t\omega_1}) \int_H |D\varphi|^2 d\mu.$$

Finally, letting $t$ tend to $+\infty$, and recalling (2.36), gives

$$\|\varphi\|_{L^2(H,\mu)}^2 \log(\|\varphi\|_{L^2(H,\mu)}^2) - \int_H \varphi^2 \log(\varphi^2) d\mu \geq -\frac{1}{\omega_1} \int_H |D\varphi|^2 d\mu$$

and the conclusion follows. □

## 2.7.1 Hypercontractivity of $R_t$

We assume here that Hypothesis 2.1 holds with $M = 1, \omega < 0$ and $C = I$ so that the semigroup $R_t$ is symmetric by Proposition 2.47. We show now that $R_t$ is hypercontractive, see [66].

**Theorem 2.56.** *For all $t > 0$ we have*

$$\|R_t\varphi\|_{L^{q(t)}(H,\nu)} \leq \|\varphi\|_{L^p(H,\nu)}, \quad p \geq 2, \; \varphi \in L^p(H,\nu), \tag{2.61}$$

*where*

$$q(t) = 1 + (p-1)e^{2\omega t}, \quad t > 0. \tag{2.62}$$

*Proof.* It is enough to show (2.61) for $\varphi \geq \varepsilon > 0$ and $\varphi \in \mathscr{E}_A(H)$. We set

$$G(t) = \|R_t\varphi\|_{L^{q(t)}(H,\nu)}, \quad F(t) = G(t)^{q(t)} = \int_H (R_t\varphi)^{q(t)} d\nu.$$

We are going to show that

$$G'(t) \leq 0. \tag{2.63}$$

This will imply $G(t) \leq G(1)$, that coincides with (2.61). Since

$$G'(t) = G(t) \left( -\frac{q'(t)}{q^2(t)} \log F(t) + \frac{1}{q(t)} \frac{F'(t)}{F(t)} \right),$$

it is enough to show that

$$-\frac{1}{q(t)} F(t) \log F(t) + \frac{F'(t)}{q'(t)} \leq 0. \tag{2.64}$$

Notice now that

$$F'(t) = \int_H (R_t\varphi)^{q(t)} q'(t) \log(R_t\varphi) d\nu + q(t) \int_H (R_t\varphi)^{q(t)-1} L_2 R_t\varphi d\nu. \tag{2.65}$$

Setting $f = (R_t\varphi)^{\frac{q(t)}{2}}$ and using the basic integration by parts inequality (2.41), we find that

$$\frac{F'(t)}{q'(t)} = \int_H f^2 \log(f^{\frac{2}{q(t)}}) d\nu + \frac{q(t)}{q'(t)} \int_H f^{2\frac{q(t)-1}{q(t)}} L_2 \left( f^{\frac{2}{q(t)}} \right) d\nu$$

$$= \frac{1}{q(t)} \int_H f^2 \log(f^2) d\nu - \frac{q(t)}{2q'(t)} \int_H \left\langle D\left( f^{\frac{2}{q(t)}} \right), D\left( f^{2\frac{q(t)-1}{q(t)}} \right) \right\rangle d\nu$$

$$= \frac{1}{q(t)} \int_H f^2 \log(f^2) d\nu - 2\frac{q(t)-1}{q'(t)q^2(t)} \int_H |Df|^2 d\nu.$$

Consequently (2.64) is equivalent to

$$\int_H f^2 \log(f^2) d\nu \leq 2 \frac{q(t)-1}{q'(t)} \int_H |Df|^2 d\nu + \overline{f^2} \log(\overline{f^2}). \tag{2.66}$$

Since

$$2 \frac{q(t)-1}{q'(t)q(t)} = \frac{1}{\omega},$$

we see that (2.66) holds in view of the log-Sobolev inequality (2.59). $\qquad\square$

**Remark 2.57.** Other proofs of hypercontractivity of $R_t$ (without assuming symmetry of $R_t$) can be found in [63] where an argument due to Neveu [85] was generalized and in [23] where the second quantization operator was used.

## 2.8 Some complements

### 2.8.1 Further regularity results when $R_t$ is strong Feller

Here we assume that Hypotheses 2.1 (with $\omega < 0$), 2.24 and 2.26 hold. Thus, the semigroup $R_t$ is strong Feller (see Remark 2.25). We have seen in §2.6 that in this case $R_t$ has a smoothing effect in the space $C_b(H)$. We want to show now that a similar smoothing effect holds in $L^2(H, \mu)$ as well.

**Proposition 2.58.** *For all $t > 0$, $\varphi \in L^2(H, \mu)$, we have $R_t \varphi \in W^{1,2}(H, \mu)$ and*

$$\|DR_t\varphi\|_{L^2(H,\mu)} \leq \|\Lambda_t\| \|\varphi\|_{L^2(H,\mu)}, \tag{2.67}$$

*where $\Lambda_t = Q_t^{-1/2} e^{tA}$.*

*Proof.* Since $C_b(H)$ is dense in $L^2(H, \mu)$, it is enough to show (2.67) for $\varphi \in C_b(H)$. Let $\varphi \in C_b(H)$. Then by (2.33) we have

$$|DR_t\varphi(x)|^2 \leq \|\Lambda_t\|^2 R_t(\varphi^2), \quad t > 0, x \in H.$$

Integrating this inequality over $H$ with respect to $\mu$, and taking into account the invariance of $\mu$, yields

$$\int_H |DR_t\varphi(x)|^2 \mu(dx) \leq \|\Lambda_t\|^2 \int_H |\varphi(x)|^2 \mu(dx), \quad t > 0. \qquad \square$$

**Proposition 2.59.** *For all $f \in L^2(H, \mu)$, we have $R(\lambda, L_2)f \in W^{1,2}(H, \mu)$ and*

$$\|DR(\lambda, L_2)f\|_{L^2(H,\mu)} \leq \gamma(\lambda)\|f\|_{L^2(H,\mu)},$$

*where $\gamma$ is defined by (2.30).*
*Moreover*

$$D(L_2) \subset W^{1,2}(H, \mu), \tag{2.68}$$

*with continuous embedding.*

*Proof.* The first statement follows from Proposition 2.58 by using Laplace transform. Let us prove (2.68). Let $\varphi \in D(L_2)$ and set $f = R(1, L_2)\varphi$ so that $\varphi - L_2\varphi = f$. Then $\varphi \in W^{1,2}(H, \mu)$ by the first statement. $\qquad \square$

### 2.8.2 The case when $A$ and $C$ commute

In this section we shall assume, besides Hypothesis 2.1 that for all $t > 0$ we have $Ce^{tA} = e^{tA}C$ for all $t \geq 0$; we shall say that $A$ and $C$ *commute*.

In this case, even if the semigroup $R_t$ is not strong Feller, it possesses smoothing properties in the directions of $C^{1/2}(H)$. We recall that this happens in the case of the Gross Laplacian, see e.g. [51, Chapter 2].

Let us introduce the derivative $D_C$ in the directions of $C^{1/2}(H)$. Formally it is given by $D_C\varphi = C^{1/2}D\varphi$.

We say that $\varphi \in C_C^1(H)$ if:

(i) There exists the limit

$$\lim_{t\to 0} \frac{1}{t} \left(\varphi(x + tC^{1/2}h) - \varphi(x)\right) := \langle D_C\varphi(x), h\rangle.$$

(ii) $D_C\varphi \in C_b^1(H; H)$.

We set

$$Q_{1,t}x = \int_0^t e^{sA}e^{sA^*}x\, ds, \quad x \in H,$$

so that $Q_t = CQ_{1,t}$, and

$$\Lambda_{1,t} = Q_{1,t}^{-1/2}e^{tA}, \quad t > 0.$$

Notice that

$$Q_t^{-1/2}e^{tA}C^{1/2} = \Lambda_{1,t}$$

and

$$\|\Lambda_{1,t}\| \le \frac{1}{\sqrt{t}} \sup_{s\in[0,t]} e^{2\omega s}, \quad t \ge 0,$$

see Remark 2.25.

The following two propositions can be proved by arguing as for the proofs of Proposition 2.28 and 2.29 respectively.

**Proposition 2.60.** *Assume that Hypothesis 2.1 holds and that $A$ and $C$ commute. Let $k \in \mathbb{N} \cup \{0\}$ and $\varphi \in C_b(H)$. Then if $t > 0$ we have $R_t\varphi \in C_C^1(H)$ and*

$$\langle D_C R_t\varphi(x), h\rangle = \int_H \langle \Lambda_{1,t}h, Q_t^{-1/2}y\rangle \varphi(e^{tA}x + y) N_{Q_t}(dy),$$

*for all $h \in H$. Moreover,*

$$\|D_C R_t\varphi\|_0 \le \frac{1}{\sqrt{t}} \sup_{s\in[0,t]} e^{2\omega s}\|\varphi\|_0, \quad t > 0.$$

**Proposition 2.61.** *Assume that Hypothesis 2.1 holds and that $A$ and $C$ commute. Let $f \in C_b(H)$ and $\lambda > 0$. Then $\varphi = R(\lambda, L)f \in C_C^1(H)$ and*

$$\|D_C\varphi\|_0 \le \sqrt{\frac{\pi}{\lambda}}\, \|f\|_0.$$

*Moreover, $D(L) \subset C_C^1(H)$ with continuous embedding.*

### 2.8.3 The Ornstein–Uhlenbeck semigroup in the space of functions of quadratic growth

We are here concerned with the *Ornstein–Uhlenbeck semigroup* acting on the space $C_{b,2}(H)$ (defined in §1.1.2).

We will need a simple generalization of Proposition 1.3 (with $\mathscr{E}_A(H)$ replacing $\mathscr{E}(H)$)

**Proposition 2.62.** *For all* $\varphi \in C_{b,2}(H)$ *there exists a three-index sequence* $\{\varphi_{n_1,n_2,n_3}\} \subset \mathscr{E}_A(H)$ *such that*

(i) $\|\varphi_{n_1,n_2,n_3}\|_{b,2} \le \|\varphi\|_{b,2}, \quad n_1, n_2, n_3 \in \mathbb{N}$.

(ii) $\lim\limits_{n_1 \to \infty} \lim\limits_{n_2 \to \infty} \lim\limits_{n_3 \to \infty} \varphi_{n_1,n_2,n_3}(x) = \varphi(x), \quad x \in H$.

*Proof.* Let $\{\varphi_{n_1,n_2}\} \subset \mathscr{E}(H)$ be a two–index sequence fulfilling the conditions of Proposition 1.3. Then, it is enough to set

$$\varphi_{n_1,n_2,n_3}(x) = \varphi_{n_1,n_2}(n_3(n_3 - A)^{-1}(x)), \quad x \in H, \; n_1, n_2, n_3 \in \mathbb{N}. \qquad \square$$

**Proposition 2.63.** *Assume that Hypothesis 2.1 holds.*

(i) *For all* $t \ge 0$, $R_t$ *is a linear bounded operator from* $C_{b,2}(H)$ *into itself. Moreover,*

$$\|R_t \varphi\|_{b,2} \le (\max\{1, M^2 e^{2\omega t}\} + \operatorname{Tr} Q_t)\|\varphi\|_{b,2}, \quad t \ge 0. \tag{2.69}$$

(ii) *If* $\varphi \in C_{b,2}(H)$ *and* $\{\varphi_{n_1,n_2,n_3}\} \subset C_{b,2}(H)$ *is a three-index sequence such that*

$$\lim\limits_{n_1 \to \infty} \lim\limits_{n_2 \to \infty} \lim\limits_{n_3 \to \infty} \varphi_{n_1,n_2,n_3}(x) = \varphi(x), \; x \in H$$

*and*

$$\|\varphi_{n_1,n_2,n_3}\|_{0,2} \le \|\varphi\|_{0,2}, \quad n_1, n_2, n_3 \in \mathbb{N},$$

*then*

$$\lim\limits_{n_1 \to \infty} \lim\limits_{n_2 \to \infty} \lim\limits_{n_3 \to \infty} R_t \varphi_{n_1,n_2,n_3}(x) = R_t \varphi(x), \quad x \in H, \; t \ge 0.$$

*Proof.* (i) Let us first check that $C_{b,2}(H)$ is stable for $R_t$. In fact if $\varphi \in C_{b,2}(H)$, we have

$$\frac{|R_t \varphi(x)|}{1 + |x|^2} \le \|\varphi\|_{0,2} \int_H \frac{1 + |e^{tA}x + y|^2}{1 + |x|^2} \, N_{Q_t}(dy).$$

Taking into account that

$$\int_H \langle e^{tA}x, y \rangle N_{Q_t}(dy) = 0,$$

we have

$$\frac{|R_t\varphi(x)|}{1+|x|^2} \le \|\varphi\|_{0,2} \int_H \frac{1+|e^{tA}x|^2+|y|^2}{1+|x|^2} \, N_{Q_t}(dy)$$

$$\le \|\varphi\|_{0,2} \left( \max\{1, M^2 e^{2\omega t}\} + \int_H |y|^2 N_{Q_t}(dy) \right)$$

$$\le \|\varphi\|_{0,2} \left( \max\{1, M^2 e^{2\omega t}\} + \mathrm{Tr}Q_t \right).$$

So, $R_t\varphi \in C_{b,2}(H)$ and (2.69) holds. The other properties are easy to check. $\qquad\square$

By Proposition 2.63 it follows that $R_t$ is a semigroup of linear bounded operators on $C_{b,2}(H)$ (not strongly continuous unless $A = 0$).

In a similar way, recalling (2.23), we can prove the following result.

**Proposition 2.64.** *Assume that Hypothesis 2.1 holds. For all $t \ge 0$, $R_t$ is a linear bounded operator from $C_{b,2}^1(H)$ into itself. Moreover*

$$|DR_t\varphi(x)| \le Me^{\omega t} \left(\max\{1, M^2 e^{2\omega t}\} + \mathrm{Tr}\, Q_t\right)\|\varphi\|_{1,2}(1+|x|^2), \ t \ge 0, \ x \in H. \tag{2.70}$$

We now introduce the infinitesimal generator of $R_t$ in $C_{b,2}(H)$. Proceeding as in §2.3.2 we define

$$(D(L), C_{b,2}(H)) = \left\{ \varphi \in C_{b,2}(H) : \lim_{h\to 0} \frac{1}{h} \left(R_h\varphi(x) - \varphi(x)\right) = f(x), \quad x \in H, \right.$$

$$\left. \sup_{h\in(0,1]} \frac{1}{h} \|R_h\varphi - \varphi\|_{0,2} < +\infty \right\},$$

and

$$L\varphi(x) = \lim_{h\to 0} \frac{1}{h} \left(R_h\varphi(x) - \varphi(x)\right), \quad x \in H, \ \varphi \in (D(L), C_{b,2}(H)).$$

The following result can be proved as in Proposition 2.21.

**Proposition 2.65.** *Assume that Hypothesis 2.1 holds.*

(i) *We have $(\max\{0, 2\omega\}, +\infty) \in \rho(L)$, and*

$$R(\lambda, L)f(x) = \int_0^{+\infty} e^{-\lambda t} R_t f(x) dt, \quad x \in H, \ f \in C_{b,2}(H). \tag{2.71}$$

*Moreover, for all $f \in C_{b,2}(H)$, we have:*

$$\|R(\lambda, L)f\|_{0,2} \le \left( \max\left\{\frac{1}{\lambda}, \frac{M^2}{\lambda - 2\omega}\right\} + \frac{1}{\lambda} \, \mathrm{Tr}\, Q_\infty \right) \|f\|_{0,2}.$$

(ii) *If* $f \in C_{b,2}(H)$ $\{f_{n_1,n_2,n_3}\}$ *is a three-index sequence such that*

$$\lim_{n_1 \to \infty} \lim_{n_2 \to \infty} \lim_{n_3 \to \infty} f_{n_1,n_2,n_3}(x) = f(x), \ x \in H,$$

*and* $\|f_{n_1,n_2,n_3}\|_{0,2} \le \|f\|_{0,2}$ *for all* $n_1, n_2, n_3 \in \mathbb{N}$, *then*

$$\lim_{n_1 \to \infty} \lim_{n_2 \to \infty} \lim_{n_3 \to \infty} R(\lambda, L) f_{n_1,n_2,n_3}(x) = R(\lambda, L) f(x),$$

*for all* $x \in H$, $\lambda > 0$

**Remark 2.66.** If $h \in D(A^*)$ we have $\varphi_h \in (D(L), C_{b,2}(H))$ [7] and

$$L\varphi_h = \frac{1}{2} \operatorname{Tr} [CD^2\varphi_h] + \langle x, A^* D\varphi_h \rangle.$$

In fact, recalling that $R_t \varphi_h(x) = e^{-\frac{1}{2} \langle Q_t x, x \rangle} e^{i \langle x, e^{tA^*} h \rangle}$, we have

$$\lim_{t \to 0} \frac{1}{t} \left( R_t \varphi_h(x) - \varphi_h(x) \right) = \left( -\frac{1}{2} \langle Ch, h \rangle + i\langle x, A^* h \rangle \right) \varphi_h(x)$$

$$= \frac{1}{2} \operatorname{Tr} [CD^2\varphi_h(x)] + \langle x, A^* \varphi_h(x) \rangle.$$

Moreover

$$\left| \frac{1}{t} \left( R_t \varphi_h(x) - \varphi_h(x) \right) \right| \le \frac{1}{t} \left| e^{-\frac{1}{2} \langle Q_t h, h \rangle} - 1 \right| + \frac{1}{t} \left| e^{i \langle e^{tA^*} h, x \rangle} - e^{i \langle h, x \rangle} \right|$$

$$\le \frac{1}{2t} \langle Q_t h, h \rangle + \frac{1}{t} \left| \langle e^{tA^*} h - h, x \rangle \right|$$

$$\le \frac{M^2}{2t} \int_0^t e^{2\omega s} ds \, |h|^2 + M \int_0^t e^{\omega s} ds + |x| \, |A^* h|.$$

Therefore

$$\sup_{t \in (0,1]} \frac{1}{t} \|R_h \varphi_h - \varphi_h\|_{0,2} < +\infty,$$

and so $\varphi_h \in (D(L), C_{b,2}(H))$ as claimed.

Notice that if $h \notin D(A^*)$ we have $\varphi_h \notin D(L)$.

By Remark 2.66 it follows that $\mathscr{E}_A(H) \subset (D(L), C_{b,2}(H))$ and

$$L\varphi = \frac{1}{2} \operatorname{Tr} [CD^2\varphi] + \langle x, A^* D\varphi \rangle, \text{ for all } \varphi \in C_{b,2}(H).$$

We want now to show, following [48], that $\mathscr{E}_A(H)$ is (in a generalized meaning), a core for $(D(L), C_{b,2}(H))$.

---

[7] recall that $\varphi_h(x) = e^{i \langle h, x \rangle}$, $x \in H$.

**Proposition 2.67.** *Assume that Hypothesis 2.1 holds and let* $\varphi \in (D(L), C_{b,2}(H))$. *Then there exists a three-index sequence* $\{\varphi_{n_1,n_2,n_3}\} \subset \mathscr{E}_A(H)$ *such that:*

(i) $\lim\limits_{n_1 \to \infty} \lim\limits_{n_2 \to \infty} \lim\limits_{n_3 \to \infty} \varphi_{n_1,n_2,n_3}(x) = \varphi(x)$, $x \in H$.

(ii) $\lim\limits_{n_1 \to \infty} \lim\limits_{n_2 \to \infty} \lim\limits_{n_3 \to \infty} L\varphi_{n_1,n_2,n_3}(x) = L\varphi(x)$, $x \in H$.

(iii) $\sup\limits_{n_1,n_2,n_3 \in \mathbb{N}} (\|\varphi_{n_1,n_2,n_3}\|_{0,2} + \|L\varphi_{n_1,n_2,n_3}\|_{0,2}) < +\infty$.

*Proof.* Let $\varphi \in (D(L), C_{b,2}(H))$ and $\lambda \in \rho(L)$. Set $\lambda\varphi - L\varphi = f$, so that $\varphi = R(\lambda, L)f$. By Proposition 2.62 there exists a three-index sequence $\{f_{n_1,n_2,n_3}\} \subset \mathscr{E}_A(H)$ such that

$$f_{n_1,n_2,n_3}(x) \to f(x) \text{ for all } x \in H, \quad \|f_{n_1,n_2,n_3}\|_{0,2} \le \|f\|_{0,2} \text{ for all } n_1, n_2, n_3 \in \mathbb{N}.$$

Set moreover $\psi_{n_1,n_2,n_3} = R(\lambda, L)f_{n_1,n_2,n_3}$. Then by Proposition 2.65 we have, for a suitable constant $C_\lambda$, that

$$\psi_{n_1,n_2,n_3}(x) \to \varphi(x), \ L\psi_{n_1,n_2,n_3}(x) \to L\varphi(x) \text{ for all } x \in H,$$

$$\|\psi_{n_1,n_2,n_3}\|_{0,2} \le C_\lambda \|f\|_{0,2}, \ \|L\psi_{n_1,n_2,n_3}\|_{0,2} \le C_\lambda \|f\|_{0,2}, \text{ for all } n_1, n_2, n_3 \in \mathbb{N}.$$

Now let $Y$ be the closure of $\mathscr{E}_A(H)$ in $C_{b,2}(H)$. Obviously $Y$ is invariant for $R_t$.

*Step* 1. The restriction of $R_t$ to $Y$ is a strongly continuous semigroup.

We have in fact, recalling that $R_t\varphi_h(x) = e^{-\frac{1}{2}\langle Q_t x, x\rangle} e^{i\langle x, e^{tA^*}h\rangle}$,

$$|R_t\varphi_h(x) - \varphi_h(x)| \le \left|e^{-\frac{1}{2}\langle Q_t h, h\rangle} - 1\right| + \left|e^{i\langle e^{tA^*}h, x\rangle} - e^{i\langle h, x\rangle}\right|$$

$$\le \frac{1}{2}\langle Q_t h, h\rangle + \left|\langle e^{tA^*}h - h, x\rangle\right|.$$

Consequently

$$\frac{|R_t\varphi_h(x) - \varphi_h(x)|}{1 + |x|^2} \le \frac{\left|e^{-\frac{1}{2}\langle Q_t h, h\rangle} - 1\right|}{1 + |x|^2} + \frac{\left|\langle e^{tA^*}h - h, x\rangle\right|}{1 + |x|^2}.$$

Therefore

$$\lim_{t \to 0} \frac{|R_t\varphi_h(x) - \varphi_h(x)|}{1 + |x|^2} = 0, \text{ uniformly in } x \in H$$

and so,

$$\lim_{t \to 0} R_t\varphi_h = \varphi_h \text{ in } C_{b,2}(H).$$

Step 1 is proved.

Let us denote by $L^Y$ the infinitesimal generator of the restriction of $R_t$ to $Y$. Its domain is given by

$$D(L^Y) = \{\varphi \in Y \cap D(L, C_{b,2}(H)) : L\varphi \in Y\}.$$

*Step 2.* $\mathscr{E}_A(H) \subset D(L^Y)$.

It is enough to check that $L\varphi_h \in Y$ for all $h \in D(A^*)$. Since

$$L\varphi_h(x) = \left[-\frac{1}{2}\langle Ch, h\rangle + i\langle x, A^*h\rangle\right],$$

it is enough to show that for all $k \in H$ the function $\psi(x) = \langle x, k\rangle$ belongs to $Y$. Set

$$\psi_\varepsilon(x) = \frac{1}{\varepsilon}\sin[\varepsilon\langle x, k\rangle], \quad \varepsilon > 0, x \in H.$$

Then there exists $\kappa > 0$ such that

$$|\psi - \psi_\varepsilon(x)| \le \varepsilon\kappa|\langle x, k\rangle|^2,$$

so that $\psi_\varepsilon \to \psi$ in $C_{b,2}(H)$ and the claim is proved.

Now we can conclude the proof. By a classical result, $\mathscr{E}_A(H)$ is a core for $L^Y$ since it is invariant for $R_t$, see [52]. Thus any function $\psi_{n_1,n_2,n_3}$ of the previous sequence can be approximated uniformly (in the graph norm of $L$) by elements on $\mathscr{E}_A(H)$ and the conclusion follows. $\qquad\square$

In a similar way one can prove the following result.

**Proposition 2.68.** *Let $\varphi \in (D(L), C_{b,2}(H)) \cap C^1_{b,2}(H)$ such that $L\varphi \in C^1_{b,2}(H)$. Then there exists a three-index sequence $\{\varphi_{n_1,n_2,n_3}\} \subset \mathscr{E}_A(H)$ such that:*

(i) $\lim\limits_{n_1\to\infty}\lim\limits_{n_2\to\infty}\lim\limits_{n_3\to\infty}\varphi_{n_1,n_2,n_3}(x) = \varphi(x)$,

$\lim\limits_{n_1\to\infty}\lim\limits_{n_2\to\infty}\lim\limits_{n_3\to\infty}L\varphi_{n_1,n_2,n_3}(x) = L\varphi(x)$ *for all $x \in H$.*

(ii) $\lim\limits_{n_1\to\infty}\lim\limits_{n_2\to\infty}\lim\limits_{n_3\to\infty}D\varphi_{n_1,n_2,n_3}(x) = D\varphi(x)$ *for all $x \in H$.*

(iii) $\sup\limits_{n_1,n_2,n_3\in\mathbb{N}}(\|\varphi_{n_1,n_2,n_3}\|_{1,2} + \|D\varphi_{n_1,n_2,n_3}\|_{1,2} + \|L\varphi_{n_1,n_2,n_3}\|_{1,2}) < +\infty.$

We now assume that Hypotheses 2.1 and 2.24 hold, so that $R_t$ is strong Feller.

The following result can be proved as Proposition 2.28.

**Proposition 2.69.** *Assume that Hypotheses 2.1 and 2.24 hold. Let $\varphi \in C_{b,2}(H)$. Then for all $t > 0$ we have $R_t\varphi \in C^1_{b,2}(H)$ and (2.31) holds. Moreover,*

$$|DR_t\varphi(x)| \le \|\Lambda_t\|(\max\{1, M^2e^{2\omega t}\} + 3\mathrm{Tr}\, Q_t)(1 + |x|^2)\|\varphi\|_{0,2}.$$

**Proposition 2.70.** *Assume that Hypotheses 2.1, 2.24 and 2.26 hold. Let $f \in C_{b,2}(H)$. Then $\varphi = R(\lambda, L)f \in C^1_{b,2}(H)$ and there exists an analytic function $\gamma_1(\lambda) \to 0$ as $\lambda \to +\infty$ such that*

$$|D\varphi(x)| \leq \gamma_1(\lambda) \, \|f\|_{b,2}.$$

*Moreover $(D(L), C_{b,2}(H)) \subset C^1_{b,2}(H)$.*

The following result can be proved as Corollary 2.30.

**Corollary 2.71.** *Assume that Hypotheses 2.1, 2.24 and 2.26 hold and that $\{\varphi_{n_1,n_2,n_3}\} \subset C_{b,2}(H)$ is a three-index sequence such that:*

(i) $\displaystyle \lim_{n_1 \to \infty} \lim_{n_2 \to \infty} \lim_{n_3 \to \infty} \varphi_{n_1,n_2,n_3}(x) = \varphi(x), \quad x \in H.$

$\displaystyle \lim_{n_1 \to \infty} \lim_{n_2 \to \infty} \lim_{n_3 \to \infty} L\varphi_{n_1,n_2,n_3}(x) = L\varphi(x), \quad x \in H.$

(ii) $\displaystyle \sup_{n_1,n_2,n_3} \{\|\varphi_{n_1,n_2,n_3}\|_{0,2} + \|L\varphi_{n_1,n_2,n_3}\|_{0,2}\} < +\infty.$

*Then we have*

$$\lim_{n_1 \to \infty} \lim_{n_2 \to \infty} \lim_{n_3 \to \infty} D\varphi_{n_1,n_2,n_3}(x) = D\varphi(x), \quad x \in H,$$

*and*

$$\sup_{n_1,n_2,n_3} \{\|D\varphi_{n_1,n_2,n_3}\|_{1,2}\} < +\infty.$$

From Proposition 2.67 and Corollary 2.71 the result follows immediately.

**Proposition 2.72.** *Assume that Hypotheses 2.1, 2.24 and 2.26 hold and let $\varphi \in (D(L), C_{b,2}(H))$. Then there exists a three-index sequence $\{\varphi_{n_1,n_2,n_3}\} \subset \mathscr{E}_A(H)$ such that:*

(i) $\displaystyle \lim_{n_1 \to \infty} \lim_{n_2 \to \infty} \lim_{n_3 \to \infty} \varphi_{n_1,n_2,n_3}(x) = \varphi(x), \quad x \in H.$

$\displaystyle \lim_{n_1 \to \infty} \lim_{n_2 \to \infty} \lim_{n_3 \to \infty} L\varphi_{n_1,n_2,n_3}(x) = L\varphi(x), \quad x \in H.$

$\displaystyle \lim_{n_1 \to \infty} \lim_{n_2 \to \infty} \lim_{n_3 \to \infty} D\varphi_{n_1,n_2,n_3}(x) = D\varphi(x), \quad x \in H.$

(ii) $\displaystyle \sup_{n_1,n_2,n_3} (\|\varphi_{n_1,n_2,n_3}\|_{1,2} + \|L\varphi_{n_1,n_2,n_3}\|_{1,2} + \|D\varphi_{n_1,n_2,n_3}\|_{1,2}) < +\infty.$

We end this section proving that $(D(L), C_b(H))$ is an algebra.

**Proposition 2.73.** *Assume that Hypotheses 2.1, 2.24 and 2.26 hold. Then if $\varphi \in (D(L), C_b(H))$, we have $\varphi^2 \in (D(L), C_b(H))$ and*

$$L(\varphi^2) = 2\varphi L\varphi + |C^{1/2}D\varphi|^2. \tag{2.72}$$

*Proof.* Let $\varphi \in (D(L), C_b(H))$, then we have obviously $\varphi \in (D(L), C_{b,2}(H))$. Let $\{\varphi_{n_1,n_2,n_3}\} \subset \mathscr{E}_A(H)$ be a three-index sequence fulfilling the conditions (i) and (ii) of Proposition 2.72. By a simple verification we have

$$L(\varphi_{n_1,n_2,n_3}^2) = 2\varphi_{n_1,n_2,n_3} L\varphi_{n_1,n_2,n_3} + |C^{1/2}D\varphi_{n_1,n_2,n_3}|^2, \quad n_1, n_2, n_3 \in \mathbb{N}.$$

Then, letting $n_3 \to \infty$, $n_2 \to \infty$, $n_1 \to \infty$ yields (2.72). $\qquad\square$

# Chapter 3

# Stochastic Differential Equations with Lipschitz Nonlinearities

## 3.1 Introduction and setting of the problem

As in Chapter 2, we are given two separable Hilbert spaces $H$ and $U$ and two linear operators $A \colon D(A) \subset H \to H$ and $B \in L(U)$ fulfilling Hypothesis 2.1. We shall denote by $M > 0$ and $\omega \in \mathbb{R}$ numbers such that

$$\|e^{tA}\| \leq M e^{\omega t}, \quad t \geq 0.$$

Moreover, for any $T > 0$ we set

$$M_T = \sup_{t \in [0,T]} \|e^{tA}\|.$$

We consider a cylindrical Wiener process $W$ defined formally as before by

$$W(t) = \sum_{k=1}^{\infty} \beta_k(t) e_k, \quad t \geq 0,$$

where $\{e_k\}$ is a complete orthonormal basis in $U$ and $\{\beta_k\}$ a sequence of mutually independent standard Brownian motions on a probability space $(\Omega, \mathscr{F}, \mathbb{P})$ given once and for all.

We are given in addition a nonlinear (Lipschitz continuous) function $F \colon H \to H$ such that

**Hypothesis 3.1.** *There exists $K > 0$ such that*

$$|F(x) - F(y)| \leq K|x - y|, \quad x, y \in H.$$

We are concerned with the stochastic differential equation

$$\begin{cases} dX(t) = (AX(t) + F(X(t))dt + BdW(t), \\ X(0) = x \in H. \end{cases} \tag{3.1}$$

By a *mild* solution of problem (3.1) on $[0,T]$ we mean a stochastic process $X \in C_W([0,T]; H)$ such that

$$X(t) = e^{tA}x + \int_0^t e^{(t-s)A}F(X(s))ds + W_A(t), \tag{3.2}$$

where $W_A$ is the stochastic convolution defined by (2.3) [1].

To study equation (3.2) we shall use the stochastic calculus. We assume that the reader has a good knowledge of this theory when $H$ and $U$ are finite dimensional, see e.g. [70], [82].

The main problems which arise in the infinite dimensional case are due to the unboundedness of the operator $A$ and to the fact that the operator $C = BB^*$ is not necessarily of trace class. If $A \in L(H)$ and $C \in L_1^+(H)$, problem (3.1) can be treated as in the finite dimensional case. We will refer to this situation as to the *classical* case.

In §3.2 we shall prove existence and uniqueness of a mild solution $X(\cdot, x)$ of problem (3.1). Then we shall show that $X(\cdot, x)$ can be approximated by solutions of classical problems. This will be done in two steps.

*Step 1.* When $C$ is not of trace class (but $A$ is possibly unbounded), we approximate $W(\cdot)$ with a finite dimensional noise

$$W_n(t) = \sum_{k=1}^n \beta_k(t)e_k, \quad t \geq 0.$$

Then we solve the problem

$$\begin{cases} dX_n(t) = (AX_n(t) + F(X_n(t))dt + BdW_n(t), \\ X_n(0) = x, \end{cases} \tag{3.3}$$

and prove that $\lim_{n \to \infty} X_n = X$ in $C_W([0,T]; H)$.

*Step 2.* We consider the *Yosida approximations* of $A$ given by

$$A_k = kA(k - A)^{-1}, \quad k \in \mathbb{N}$$

---

[1] For the definition of the space $C_W([0,T]; H)$ see §2.2.

and the approximating problem

$$
\begin{cases}
dX_{n,k}(t) = (A_k X_{n,k}(t) + F(X_{n,k}(t))dt + BdW_n(t), \\
X_{n,k}(0) = x,
\end{cases}
\tag{3.4}
$$

Then we prove that, for any $n \in \mathbb{N}$, $\lim_{k \to \infty} X_{n,k} = X_n$ in $C_W([0,T]; H)$.

§3.3 is devoted to showing several properties of the transition semigroup

$$
P_t \varphi(x) = \mathbb{E}\left[\varphi(X(t,x))\right], \quad \varphi \in B_b(H), \ t \geq 0, \ x \in H,
\tag{3.5}
$$

as Feller, strong Feller and Markov property.

§3.4 is devoted to irreducibility of $P_t$, §3.5 to the existence and uniqueness of an invariant measure $\nu$. In §3.6 we extend the semigroup $P_t$ to $L^2(H, \nu)$ and prove that the space $\mathscr{E}_A(H)$ of exponential functions is a core for $P_t$; then in §3.7 we prove the "carré du champs identity" and deduce some consequences as Poincaré and log-Sobolev inequalities. Finally, §3.8 is devoted to compare (when $A$ is of negative type) the measure $\nu$ with the invariant measure of the Ornstein–Uhlenbeck process corresponding to the case $F = 0$.

## 3.2 Existence, uniqueness and approximation

We start with the following existence and uniqueness result.

**Theorem 3.2.** *Assume that Hypotheses 2.1 and 3.1 hold. Then for any $x \in H$ problem (3.1) has a unique mild solution $X \in C_W([0,T]; H)$.*

*Proof.* Setting $Z_T = C_W([0,T]; H)$, problem (3.2) is equivalent to the following equation in $Z_T$,

$$
X(t) = e^{tA}x + \gamma(X)(t) + W_A(t), \quad t \in [0,T],
$$

where

$$
\gamma(X)(t) = \int_0^t e^{(t-s)A} F(X(s))ds, \quad X \in Z_T, \quad t \in [0,T].
$$

It is easy to check that $\gamma$ maps $Z_T$ into itself and that

$$
\|\gamma(X) - \gamma(X_1)\|_{Z_T} \leq T M_T \|X - X_1\|_{Z_T}, \quad X, X_1 \in Z_T,
$$

(recall that $M_T = \sup_{t \in [0,T]} \|e^{tA}\|$). Now let $T_1 \in (0,T]$ be such that $T_1 M_T < 1$. Then $\gamma$ is a contraction on $Z_{T_1}$. Therefore, problem (3.1) has a unique mild solution on $[0, T_1]$. By a similar argument, one can show existence and uniqueness on $[T_1, 2T_1]$ and so on. $\qquad\square$

We shall denote by $X(\cdot, x)$ the mild solution to problem (3.1). Let us study the continuous dependence of $X(\cdot, x)$ by $x$.

**Proposition 3.3.** *Assume that Hypotheses 2.1 and 3.1 hold. Then for any $x, y \in H$ we have*

$$|X(t, x) - X(t, y)| \le M_T e^{TKM_T}|x - y|. \tag{3.6}$$

*Proof.* In fact, for $t \in [0, T]$, we have

$$X(t, x) - X(t, y) = e^{tA}(x - y) + \int_0^t e^{(t-s)A}[F(X(s, x)) - F(X(s, y))]ds.$$

Consequently

$$|X(t, x) - X(t, y)| \le M_T|x - y| + KM_T \int_0^t |X(s, x) - X(s, y)|ds$$

and (3.6) follows from Gronwall's lemma. $\qquad\square$

**Exercise 3.4.** Assume that Hypotheses 2.1, 3.1, and 2.6 hold. Prove, using Theorem 2.9, that for any $T > 0$ there exists $C(T, x) > 0$ such that

$$\mathbb{E}\left(\sup_{t \in [0,T]} |X(t, x)|^2\right) \le C(T, x).$$

Deduce that $X(\cdot, x)$ is $\mathbb{P}$-almost-sure continuous for any $x \in H$.

We now consider the approximating problems (3.3) and (3.4).

**Theorem 3.5.** *Assume that Hypotheses 2.1 and 3.1 hold. Let $x \in H$, $n, k \in \mathbb{N}$. Then there exist unique mild solutions $X_n$ and $X_{n,k}$ of problems (3.3) and (3.4) respectively. Moreover*

$$\lim_{n \to \infty} X_n = X \quad \text{in } C_W([0, T]; H), \tag{3.7}$$

*and for any $n \in \mathbb{N}$,*

$$\lim_{k \to \infty} X_{n,k} = X_n \quad \text{in } C_W([0, T]; H). \tag{3.8}$$

*Proof.* Existence and uniqueness of solutions of (3.3) and (3.4) follow from Theorem 3.2. Let us prove (3.7). For any $n \in \mathbb{N}$ we have

$$X(t) - X_n(t) = \int_0^t e^{(t-s)A}[F(X(s)) - F(X_n(s))]ds$$

$$+ \sum_{j=n+1}^{\infty} \int_0^t e^{(t-s)A}Be_j d\beta_j(s).$$

It follows that

$$\mathbb{E}\left(|X(t) - X_n(t)|^2\right) \le 2M_T^2 K^2 T \int_0^t \mathbb{E}\left(|X(s) - X_n(s)|^2\right) ds$$

$$+ \; 2\int_0^t \mathrm{Tr}\, [e^{sA} B(1 - \Pi_n) B^* e^{sA^*}] ds,$$

where $\Pi_n = \sum_{j=1}^n e_j \otimes e_j$. Now (3.7) follows using the Gronwall lemma.

Let us prove (3.8). For $n, k \in \mathbb{N}$ we have

$$X_n(t) - X_{n,k}(t) \;=\; \int_0^t e^{(t-s)A_k}[F(X_n(s)) - F(X_{n,k}(s))] ds$$

$$+ \; a(k) + b(n,k) + c(n,k), \tag{3.9}$$

where

$$a(k) = e^{tA} x - e^{tA_k} x,$$

$$b(n,k) = \int_0^t [e^{(t-s)A} - e^{(t-s)A_k}] F(X_n(s)) ds,$$

$$c(n,k) = \sum_{j=1}^n \int_0^t [e^{(t-s)A} - e^{(t-s)A_k}] Be_j d\beta_j.$$

Let us recall a well-known property of the Yosida approximations, namely that

$$\lim_{k \to \infty} e^{tA_k} x = e^{tA} x,$$

for all $x \in H$ uniformly in $t \in [0, T]$ for any $T > 0$.

Using this property, we find that

$$|a(k)| \le \sup_{t \in [0,T]} |e^{tA} x - e^{tA_k} x| := \alpha(k) \to 0 \text{ as } k \to \infty, \tag{3.10}$$

$$\mathbb{E}(|b(n,k)|) \le \int_0^T \mathbb{E}\left(\left| [e^{(t-s)A} - e^{(t-s)A_k}] F(X_n(s)) \right|\right) ds := \beta(n,k)$$

$$\to 0 \text{ as } k \to \infty, \tag{3.11}$$

$$\mathbb{E}(|c(n,k)|^2) \le \int_0^T \mathrm{Tr}\{[e^{sA} - e^{sA_k}] B\Pi_n B^* [e^{sA^*} - e^{sA_k^*}]\} ds := \gamma(n,k)$$

$$\to 0 \text{ as } k \to \infty. \tag{3.12}$$

Now from (3.9) we have

$$\mathbb{E}(|X_n(t) - X_{n,k}(t)|) \leq M_T K \int_0^t \mathbb{E}(|X_n(s) - X_{n,k}(s)|)ds$$

$$+ \; \alpha(k) + \beta(n,k) + [\gamma(n,k)]^{1/2}.$$

The conclusion follows from (3.10)–(3.12) and the Gronwall lemma.    □

### 3.2.1  Derivative of the solution with respect to the initial datum

Here we assume, besides Hypothesis 2.1, that $F$ fulfills a stronger assumption than 3.1, i.e. $F \in C_b^2(H; H)$ [(2)]. Our goal is to prove that in this case the mild solution $X(t,x)$ of (3.1) is differentiable with respect to $x$ and that for any $h \in H$ we have

$$DX(t,x) \cdot h = \eta^h(t,x), \quad t \geq 0, \; x \in H,$$

where $\eta^h(t,x)$ is the mild solution of the equation

$$\begin{cases} \dfrac{d}{dt} \, \eta^h(t,x) &= \; A\eta^h(t,x) + DF(X(t,x)) \cdot \eta^h(t,x), \\[2mm] \eta^h(0,x) &= \; h. \end{cases} \tag{3.13}$$

This means that $\eta^h(t,x)$ is the solution of the integral equation

$$\eta^h(t,x) = e^{tA}h + \int_0^t e^{(t-s)A} DF(X(s,x)) \cdot \eta^h(s,x)ds, \quad t \geq 0.$$

It is easy to see, proceeding as for the proof of Theorem 3.5, that (3.13) has a unique mild solution $\eta^h(t,x)$. Moreover, for any $T > 0$ we have

$$\lim_{n \to \infty} \eta_n^h = \eta^h, \quad \lim_{k \to \infty} \eta_{n,k}^h = \eta_n^h \quad \text{in } C_W([0,T]; H), \quad n \in \mathbb{N}, \tag{3.14}$$

where $\eta_n^h$ and $\eta_{n,k}^h$ are the mild solution of problems

$$\begin{cases} \dfrac{d}{dt} \, \eta_n^h(t,x) &= \; A\eta_n^h(t,x) + DF(X_n(t,x)) \cdot \eta_n^h(t,x), \\[2mm] \eta_n^h(0,x) &= \; h, \end{cases} \tag{3.15}$$

and

$$\begin{cases} \dfrac{d}{dt} \, \eta_{n,k}^h(t,x) &= \; A_k\eta_{n,k}^h(t,x) + DF(X_{n,k}(t,x)) \cdot \eta_{n,k}^h(t,x), \\[2mm] \eta_{n,k}^h(0,x) &= \; h, \end{cases} \tag{3.16}$$

respectively.

---

[(2)] It would be enough to take $F \in C_b^1(H; H)$.

**Theorem 3.6.** *Assume, besides Hypothesis 2.1, that $F \in C_b^2(H; H)$. Then $X(t, x)$ is differentiable with respect to $x$ ($\mathbb{P}$-a.s.) and for any $h \in H$ we have*

$$DX(t, x) \cdot h = \eta^h(t, x), \quad \mathbb{P}\text{-a.s.} \tag{3.17}$$

*where $\eta^h(t, x)$ is the mild solution of (3.13). Moreover,*

$$|\eta^h(t, x)| \le Me^{(\omega + MK)t}|h|, \quad t \ge 0. \tag{3.18}$$

*Finally,*

$$|\eta_{n,k}^h(t, x)| \le Me^{(\omega + MK)t}|h|, \quad t \ge 0, \ x \in H, \ n, k \in \mathbb{N}. \tag{3.19}$$

*Proof.* We first notice that if $\eta^h(t, x)$ is the mild solution of (3.13) we have

$$|\eta^h(t, x)| \le Me^{\omega t}|h| + MK \int_0^t e^{\omega(t-s)}|\eta^h(s, x)|ds,$$

so that (3.18) (as well as (3.19)) follows from the Gronwall lemma.

Let us prove now that $\eta^h(t, x)$ fulfills (3.17). For this fix $T > 0$, $x \in H$ and $h \in H$ such that $|h| \le 1$. We are going to show that there is $C_T > 0$ such that

$$|X(t + h, x) - X(t, x) - \eta^h(t, x)| \le C_T|h|^2, \quad \mathbb{P}\text{-a.s.} \tag{3.20}$$

Setting

$$r^h(t, x) = X(t, x + h) - X(t, x) - \eta^h(t, x),$$

we see that $r^h(t, x)$ satisfies the equation

$$r^h(t, x) = \int_0^t e^{(t-s)A}[F(X(s, x + h)) - F(X(s, x))]ds$$

$$- \int_0^t e^{(t-s)A}DF(X(s, x)) \cdot \eta^h(s, x)ds.$$

Consequently we have that

$$r^h(t, x) = \int_0^t e^{(t-s)A}\left[\int_0^1 DF(\rho(\xi, s))d\xi\right] \cdot (r^h(s, x) + \eta^h(s, x))ds$$

$$- \int_0^t e^{(t-s)A}DF(X(s, x)) \cdot \eta^h(s, x)ds,$$

where

$$\rho(\xi, s) = \xi X(s, x + h)) + (1 - \xi)X(s, x).$$

Therefore

$$r^h(t, x) = \int_0^t e^{(t-s)A}\left[\int_0^1 DF(\rho(\xi, s))d\xi\right] \cdot r^h(s, x)ds$$

$$+ \int_0^t e^{(t-s)A}\left[\int_0^1 (DF(\rho(\xi, s)) - DF(X(s, x)))d\xi\right] \cdot \eta^h(s, x)ds.$$

Setting $\gamma_T = \sup_{t\in[0,T]} Me^{(\omega+MK)t}$, and taking into account (3.18), we find that

$$\left|\int_0^1 (DF(\rho(\xi,s)) - DF(X(s,x)))d\xi\right|$$

$$\leq \frac{1}{2}\,\|F\|_2\,|h||X(s,x+h)) - X(s,x)|$$

$$\leq \frac{1}{2}\,\|F\|_2\,|h|\,(|r^h(s,x)| + \gamma_T\,|h|).$$

It follows that

$$|r^h(t,x)| \leq M(K + \|F\|_2)\int_0^t |r^h(s,x)|ds + TM_T\gamma_T\,\|F\|_2\,|h|^2.$$

Thus (3.20) follows from the Gronwall lemma.  $\square$

If $F \in C_b^3(H;H)$, one can prove that the mild solution $X(t,x)$ of (3.3) is twice differentiable on $x$. More precisely, proceeding as before, one can show the following result.

**Theorem 3.7.** *Assume, besides Hypothesis 2.1, that $F \in C_b^3(H;H)$. Then $X(t,x)$ is twice differentiable in $x$ and for any $h,k \in H$ we have*

$$X_{xx}(t,x) \cdot (h,k) = \zeta^{h,k}(t,x), \quad x,h \in H, \ \mathbb{P}\text{-a.s.},$$

*where $\zeta^{h,k}(t,x)$ is the mild solution to the equation*

$$\begin{cases} \dfrac{d}{dt}\,\zeta^{h,k}(t,x) &= A\zeta^{h,k}(t,x) + DF(X(t,x)) \cdot \zeta^{h,k}(t,x) \\[2mm] &\quad + D^2F(X(t,x)) \cdot (\eta^h(t,x), \eta^k(t,x)), \\[2mm] \zeta^{h,k}(0,x) &= 0. \end{cases}$$

## 3.3   The transition semigroup

Let us consider the transition semigroup corresponding to the mild solution $X(t,x)$ to (3.1), defined by

$$P_t\varphi(x) = \mathbb{E}[\varphi(X(t,x))], \quad \varphi \in B_b(H), \ t \geq 0, \ x \in H \tag{3.21}$$

and also the approximating semigroups

$$P_t^n\varphi(x) = \mathbb{E}[\varphi(X_n(t,x))], \quad \varphi \in B_b(H), \ t \geq 0, \ x \in H, \tag{3.22}$$

$$P_t^{n,k}\varphi(x) = \mathbb{E}[\varphi(X_{n,k}(t,x))], \quad \varphi \in B_b(H), \ t \geq 0, \ x \in H, \tag{3.23}$$

for all $n, k \in \mathbb{N}$, where $X_n(t, x)$ and $X_{n,k}(t, x)$ are the solutions of (3.3) and (3.4) respectively.

We have obviously for all $t \geq 0$, $n, k \in \mathbb{N}$,

$$\|P_t \varphi\|_0 \leq \|\varphi\|_0, \quad \|P_t^n \varphi\|_0 \leq \|\varphi\|_0, \quad \|P^{n,k} \varphi\|_0 \leq \|\varphi\|_0, \quad \varphi \in B_b(H).$$

Moreover by the dominated convergence theorem it follows that

$$\begin{cases} \lim_{n \to \infty} P_t^n \varphi(x) = P_t \varphi(x), & \varphi \in C_b(H), \ t \geq 0, \ x \in H, \\ \\ \lim_{k \to \infty} P_t^{n,k} \varphi(x) = P_t^n \varphi(x), & \varphi \in C_b(H), \ t \geq 0, \ x \in H, \ n \in \mathbb{N}. \end{cases} \tag{3.24}$$

If $F \in C_b^2(H; H)$, then $P_t \varphi$ is differentiable for any $t \geq 0$, in view of Theorem 3.6 and for each $h \in H$ we have

$$\langle DP_t \varphi(x), h \rangle = \mathbb{E}[\langle D\varphi(X(t, x)), \eta^h(t, x) \rangle], \quad x \in H. \tag{3.25}$$

**Exercise 3.8.** Assume, besides Hypothesis 2.1, that $F \in C_b^2(H; H)$. Show that

$$\begin{cases} \lim_{n \to \infty} DP_t^n \varphi(x) = DP_t \varphi(x), & x \in H, \ \varphi \in C_b(H), \\ \\ \lim_{k \to \infty} DP_t^{n,k} \varphi(x) = DP_t^n \varphi(x), & x \in H, \ n \in \mathbb{N}, \ \varphi \in C_b(H). \end{cases} \tag{3.26}$$

We now prove that $P_t$ fulfills the semigroup law and that it is Feller.

**Proposition 3.9.** *Assume that Hypotheses 2.1 and 3.1 hold. Then for all $t, s \geq 0$ we have*

$$P_{t+s} \varphi = P_t P_s \varphi, \quad \varphi \in B_b(H).$$

*Moreover, for all $\varphi \in C_b(H)$ and $t \geq 0$, we have $P_t \varphi \in C_b(H)$.*

*Proof.* By the Markov property for classical stochastic differential equations, the semigroup law holds for $P_t^{n,k}$. Then the conclusion follows by letting $n, k$ tend to infinity.

Let us prove the Feller property. Since $C_b^1(H)$ is dense in $C_b(H)$, see [75], it is enough to show that $P_t \varphi \in C_b(H)$ for all $\varphi \in C_b^1(H)$. For such a function $\varphi$ we have, in view of (3.6)

$$\begin{aligned} |P_t \varphi(x) - P_t \varphi(y)| &\leq \|\varphi\|_1 \mathbb{E}\left(|X(t, x) - X(t, y)|\right) \\ \\ &\leq M_T e^{TKM_T} |x - y|, \ x, y \in H. \end{aligned}$$

Therefore $P_t \varphi \in C_b(H)$ as required. $\qquad \square$

### 3.3.1 Strong Feller property

We recall that the transition semigroup $P_t$ is said to be *strong Feller* if $P_t\varphi \in C_b(H)$ for all $\varphi \in B_b(H)$ and all $t > 0$. In order to prove that $P_t$ is strong Feller we need the *Bismut–Elworthy* formula for the derivative of the transition semigroup. This formula has been proved in [10] and [58] when $H$ is finite dimensional and it has been extended to the infinite dimensional case in [41].

To prove this formula we need to approximate the Lipschitz continuous function $F$ by functions of class $C^2$. To this purpose, it is convenient to introduce an auxiliary Ornstein–Uhlenbeck semigroup,

$$U_t\varphi(x) = \int_H \varphi(e^{tS}x + y)N_{\frac{1}{2} S^{-1}(e^{2tS}-1)}(dy), \quad t > 0, \ x \in H, \tag{3.27}$$

where $S : D(S) \subset H \to H$ is a given self-adjoint negative definite operator such that $S^{-1}$ is of trace class. By Proposition 2.28 and (2.29) we know that $U_t$ is strong Feller.

Now, we introduce a regularization $F_\beta$ of $F$ by setting, for any $h \in H$,

$$\langle F_\beta(x), h \rangle = \int_H \langle F(e^{\beta S}x + y), e^{\beta S}h \rangle N_{\frac{1}{2} S^{-1}(e^{2\beta S}-1)}(dy), \quad \beta > 0. \tag{3.28}$$

It is easy to check that $F_\beta$ is Lipschitz continuous, of class $C^\infty$ and that its derivatives of all orders are bounded in $H$. Moreover, in view of Proposition 2.17–(iv), we have

$$\lim_{\beta \to 0} F_\beta(x) = F(x), \quad x \in H. \tag{3.29}$$

Let now $\beta > 0$. Then problem

$$\begin{cases} dX^\beta(t) = (AX^\beta(t) + F_\beta(X^\beta(t))dt + BdW(t), \\ \\ X^\beta(0) = x \in H, \end{cases} \tag{3.30}$$

has a unique mild solution $X^\beta(t, x)$, in view of Theorem 3.2. It is not difficult to check that

$$\lim_{\beta \to 0} X^\beta(t, x) = X(t, x), \quad x \in H, \ t > 0. \tag{3.31}$$

Let $P_t^\beta$ be the transition semigroup

$$P_t^\beta\varphi(x) = \mathbb{E}[\varphi(X^\beta(t, x))], \quad \varphi \in B_b(H).$$

Then by (3.31) it follows that

$$\lim_{\beta \to 0} P_t^\beta\varphi(x) = P_t\varphi(x), \quad x \in H, \ t > 0. \tag{3.32}$$

**Proposition 3.10.** *Assume that Hypothesis 2.1 holds, that $F$ is of class $C^2$ with bounded derivatives of the first and second order and that $B$ is continuously invertible. Then if $\varphi \in C_b^2(H)$ and $t > 0$, we have*

$$\langle DP_t\varphi(x), h\rangle = \frac{1}{t}\,\mathbb{E}\left[\varphi(X(t,x))\int_0^t \langle B^{-1}\eta^h(s,x), dW(s)\rangle\right], \qquad (3.33)$$

*for any $h \in H$.*

*Proof.* Let first $n, k \in \mathbb{N}$. Then $P_t^{n,k}\varphi \in C_b^1(H)$ and, applying Itô's formula to the function $s \mapsto P_{t-s}^{n,k}\varphi(X_{n,k}(s,x))$, we find that

$$\varphi(X_{n,k}(t,x)) = P_t^{n,k}\varphi(x) + \int_0^t \langle DP_{t-s}^{n,k}\varphi(X_{n,k}(s,x)), B_n dW(s)\rangle. \qquad (3.34)$$

Letting $n \to \infty, k \to \infty$ in (3.34), yields

$$\varphi(X(t,x)) = P_t\varphi(x) + \int_0^t \langle DP_{t-s}\varphi(X(s,x)), BdW(s)\rangle.$$

Now, multiplying both sides of this identity by

$$\int_0^t \langle B^{-1}\eta^h(s,x), dW(s)\rangle,$$

taking expectation and using Markov property [3], yields (3.33).  □

**Theorem 3.11.** *Assume that Hypotheses 2.1 and 3.1 hold and $B$ is continuously invertible. Then $P_t$ is strong Feller.*

*Proof.* Step 1. We assume in addition that $\varphi \in C_b^2(H)$.

Let $\beta > 0$ and let $F_\beta$ be defined by (3.28). Then, for any $h \in H$ we have by (3.33), using the Hölder inequality and recalling (3.19) and (3.2),

$$|\langle DP_t^\beta\varphi(x), h\rangle|^2 \leq t^{-2}\|\varphi\|_0^2\,\|B^{-1}\|_{L(H;U)}^2\,\mathbb{E}\left[\int_0^t |\eta_\beta^h(s,x)|^2 ds\right]$$

$$\leq t^{-2}\|\varphi\|_0^2\,\|B^{-1}\|_{L(H;U)}^2 M^2 t \sup_{s\in[0,t]} e^{2(\omega+K)s}\,|h|^2,$$

where $\eta_\beta^h(s,x) = DX^\beta(s,x)\cdot h$. So, by the arbitrariness of $h$ we have

$$|DP_t^\beta\varphi(x)| \leq t^{-1/2}\,\|B^{-1}\|_{L(H;U)} M \sup_{s\in[0,t]} e^{(\omega+MK)s}\,\|\varphi\|_0,$$

---

[3] The Markov property for $P_t$ follows from the Markov property of $P_t^{n,k}$.

which implies

$$|P_t^\beta \varphi(x) - P_t^\beta \varphi(y)| \le t^{-1/2} \, \|B^{-1}\|_{L(H;U)} \, M \sup_{s \in [0,t]} e^{(\omega+K)s} \, \|\varphi\|_0 \, |x - y|$$

for any $x, y \in H$.

Letting $\beta$ tend to 0 yields

$$|P_t \varphi(x) - P_t \varphi(y)| \le t^{-1/2} \, \|B^{-1}\|_{L(H;U)} \, M \sup_{s \in [0,t]} e^{(\omega+K)s} \, \|\varphi\|_0 \, |x - y|. \quad (3.35)$$

*Step* 2. Conclusion.

Fix $t > 0$, let $x, y \in H$ and let $\varphi \in B_b(H)$. We claim that (3.35) still holds. Let us consider the (signed) measure $\zeta_{x,y} = \lambda_{t,x} - \lambda_{t,y}$, where $\lambda_{t,x}$ (resp. $\lambda_{t,y}$) is the law of $X(t, x)$ (resp. $X(t, y)$) and choose a sequence $\{\varphi_n\} \subset C_b^2(H)$ such that

(i) $\displaystyle \lim_{n \to \infty} \varphi_n(z) = \varphi(z)$ for $\zeta_{x,y}$-almost $z \in H$.

(ii) $\|\varphi_n\|_0 \le \|\varphi\|_0, \quad n \in \mathbb{N}.$

Then we have

$$P_t \varphi(x) - P_t \varphi(y) = \int_H \varphi(z) \zeta_{x,y}(dx)$$

and so, by the dominated convergence theorem,

$$P_t \varphi(x) - P_t \varphi(y) = \lim_{n \to \infty} \int_H \varphi_n(z) \zeta_{x,y}(dx).$$

Now, using (3.35), we conclude that the inequality

$$|P_t \varphi(x) - P_t \varphi(y)| \le t^{-1/2} \, \|B^{-1}\|_{L(H;U)} \, M \sup_{s \in [0,t]} e^{(\omega+K)s} \, \|\varphi\|_0 \, |x - y|$$

holds for all $x, y \in H$. So, $P_t \varphi$ is continuous as required. $\qquad\square$

## 3.3.2  Irreducibility

We now study the irreducibility of $P_t$. A basic tool for proving irreducibility is the approximate controllability of the system

$$\begin{cases} y'(t) = Ay(t) + F(y(t)) + Bu(t), & t \ge 0, \\ \\ y(0) = x, \end{cases} \quad (3.36)$$

where $u \in L^2(0, T; U)$. Let us denote by $y(\cdot, x; u)$ the mild solution of (3.36), that is the solution of the integral equation

$$y(t) = e^{tA} x + \int_0^t e^{(t-s)A} F(y(s)) ds + \sigma_u(t), \quad t \ge 0, \quad (3.37)$$

where $\sigma_u$ is defined by

$$\sigma_u(t) = \int_0^t e^{(t-s)A} Bu(s)ds, \quad t \geq 0.$$

We say that system (3.36) is *approximatively controllable* in time $T > 0$ if for any $\varepsilon > 0$, $x_0, x_1 \in H$, there exists $u \in L^2(0,T;U)$ such that

$$|y(T, x_0; u) - x_1| \leq \varepsilon. \tag{3.38}$$

**Proposition 3.12.** *Assume that Hypotheses 2.1 and 3.1 hold and that $B(U)$ is dense in $H$. Then system (3.36) is approximatively controllable in any time $T > 0$.*

*Proof.* Let $T > 0$, $x_0, x_1 \in H$ and $\varepsilon > 0$ be fixed. We need the following notation. For any $z_0, z_1 \in D(A)$ we set

$$\alpha_{z_0,z_1}(t) = \frac{T-t}{T} z_0 + \frac{t}{T} z_1, \quad t \in [0,T], \tag{3.39}$$

and

$$\beta_{z_0,z_1}(t) = \frac{d}{dt} \alpha_{z_0,z_1}(t) - A\alpha_{z_0,z_1}(t) - F(\alpha_{z_0,z_1}(t)), \quad t \in [0,T]. \tag{3.40}$$

Note that
$$\alpha_{z_0,z_1}(0) = z_0, \quad \alpha_{z_0,z_1}(T) = z_1.$$

Now choose $z_0, z_1 \in D(A)$ and $u \in C([0,T]; H)$ such that

(i) $|x_0 - z_0| < c\varepsilon, \quad |x_1 - z_1| < c\varepsilon,$

(ii) $|\beta_{z_0,z_1}(t) - Bu(t)| \leq c\varepsilon, \quad t \in [0,T],$

(this is possible because $B(U)$ is dense in $H$) where $c$ is a positive constant.

We are going to show that $c$ may be chosen such as (3.37) is fulfilled. We have in fact

$$|y(T, x_0; u) - x_1| \leq |y(T, x_0; u) - y(T, z_0; u)| + |y(T, z_0; u) - x_1|$$

$$\leq |y(T, x_0; u) - y(T, z_0; u)| + |y(T, z_0; u) - \alpha_{z_0,z_1}(T)| := I_1 + I_2. \tag{3.41}$$

Let us estimate $I_1$. Set $v(t) = y(t, x_0; u) - y(t, z_0; u)$, $t \in [0,T]$. Then we have

$$v(t) = e^{tA}(x_0 - z_0) + \int_0^t e^{(t-s)A}(F(y(s, x_0; u)) - F(y(s, z_0; u)))ds, \quad t \in [0,T].$$

Consequently,

$$|v(t)| \leq M_T |x_0 - z_0| + KM_T \int_0^t |v(s)|ds, \quad t \in [0,T],$$

so that, by the Gronwall lemma,

$$|I_1| = |v(T)| \le M_T e^{K M_T T} |x_0 - z_0| \le c M_T e^{K M_T T} \varepsilon. \tag{3.42}$$

Let us estimate $I_2$. Set $z(t) = y(t, z_0; u) - \alpha_{z_0, z_1}(t)$, $t \in [0, T]$. Then we have

$$z(t) = \int_0^t e^{(t-s)A} (F(y(s, z_0; u)) - F(\alpha_{z_0, z_1}(s)))) ds$$
$$+ \int_0^t e^{(t-s)A} (Bu(s) - \beta_{z_0, z_1}(s)) ds.$$

Consequently,

$$|z(t)| \le K M_T \int_0^t |z(s)| ds + c T M_T \varepsilon, \quad t \in [0, T],$$

so that, again by the Gronwall lemma,

$$|I_2| = |z(T)| \le c T^2 M_T e^{K M_T T} \varepsilon. \tag{3.43}$$

So, by (3.41) we have, taking into account (3.42) and (3.43), that

$$|y(T, x_0; u) - x_1| \le c T M_T e^{K M_T T} \varepsilon (1 + T^2).$$

The conclusion now follows easily.                                      □

**Theorem 3.13.** *Assume that Hypotheses 2.1 and 3.1 hold and that $B(U)$ is dense in $H$. Then $P_t$ is irreducible.*

*Proof.* Fix $x_0, x_1 \in H$, $\varepsilon > 0$ and $T > 0$. Denote by $B(x_1, \varepsilon)$ the ball in $H$ with center $x_1$ and radius $\varepsilon > 0$. We wish to show that

$$P_T 1_{B(x_1, \varepsilon)}(x_0) = \mathbb{P}\left(|X(T, x_0) - x_1| < \varepsilon\right) > 0,$$

where $X(t, x_0)$ is the solution of (3.2), or equivalently that

$$P_T 1_{B(x_1, \varepsilon)}(x_0) = \mathbb{P}\left(|X(T, x_0) - x_1| \ge \varepsilon\right) \le 1. \tag{3.44}$$

By Proposition 3.12 there exists a control $u \in L^2(0, T; U)$ such that $|y(T, x_0; u) - x_1| \le \frac{\varepsilon}{2}$, where $y$ is the mild solution of (3.36). Since

$$|X(T, x_0) - x_1| \le |X(T, x_0) - y(T)| + \frac{\varepsilon}{2},$$

we have that

$$\mathbb{P}\left(|X(T, x_0) - x_1| \ge \varepsilon\right) \le \mathbb{P}\left(|X(T, x_0) - y(T)| \ge \frac{\varepsilon}{2}\right). \tag{3.45}$$

But

$$|X(t,x_0) - y(t)| \leq KM_T \int_0^t |X(s,x_0) - y(s)|ds + |W_A(t) - \sigma_A(t)|, \quad t \in [0,T].$$

So, by the Gronwall lemma, it follows that

$$|X(T,x) - y(T)| \leq e^{KM_T} \int_0^T |W_A(s) - \sigma_A(s)|ds$$

and by the Hölder inequality we deduce that

$$|X(T,x) - y(T)|^2 \leq Te^{2KM_T}\|W_A - \sigma_A\|^2_{L^2(0,T;H)}.$$

Thus by (3.45) it follows that

$$\mathbb{P}\left(|X(T,x) - z| \geq \varepsilon\right) \leq \mathbb{P}\left(\|W_A - \sigma_A\|_{L^2(0,T;H)} \geq \frac{\varepsilon}{\sqrt{T}\, e^{KM_T}}\right).$$

Now (3.44) follows since $W_A(\cdot)$ is a nondegenerate Gaussian random variable in $L^2(0,T;H)$, see Proposition 2.17. $\qquad\square$

## 3.4  Invariant measure $\nu$

We shall assume in this section that

**Hypothesis 3.14.**

  (i) *Hypothesis 2.1 holds with $M = 1$.*

 (ii) *Hypothesis 3.1 holds.*

(iii) $\omega + \kappa < 0$, *where*

$$\kappa := \inf\left\{\frac{\langle F(x) - F(y), x - y\rangle}{|x - y|^2} : x, y \in H\right\}.$$

Clearly, we have $|\kappa| \leq K$ and

$$\langle F(x) - F(y), x - y\rangle \leq \kappa|x - y|^2, \quad x, y \in H. \tag{3.46}$$

We set $\omega_1 = -\omega - \kappa$.

Under Hypothesis 3.14 we can improve the estimate (3.18) for $\eta^h(t,x)$.

**Lemma 3.15.** *Assume that, besides Hypothesis 3.14, $F \in C_b^2(H;H)$, and let $DX(t,x)\cdot h = \eta^h(t,x)$. Then we have*

$$|\eta^h(t,x)| \leq e^{-\omega_1 t}|h|, \quad t \geq 0. \tag{3.47}$$

*Proof.* Multiplying both sides of the first equation in (3.13) by $\eta^h(t,x)$ yields

$$\frac{1}{2}\frac{d}{dt}|\eta^h(t,x)|^2 = \langle A\eta^h(t,x) + DF(X(t,x))\cdot\eta^h(t,x), \eta^h(t,x)\rangle.$$

Since $M = 1$ we have

$$\langle Az, z\rangle \le \omega|z|^2, \quad \text{for all } z \in D(A).$$

Moreover, by (3.46), we have $\langle DF(X(t,x))\cdot z, z\rangle \le \kappa|z|^2$ for all $z \in H$. Consequently,

$$\frac{1}{2}\frac{d}{dt}|\eta^h(t,x)|^2 \le (\omega + \kappa)|\eta^h(t,x)|^2,$$

and the conclusion follows by a classical comparison result. $\qquad\square$

To prove the existence of an invariant measure it is useful to consider problem (3.1) with a negative initial time $s$ (see [49]), that is

$$\begin{cases} dX(t) &= [AX(t) + F(X(t))]dt + Bd\overline{W}(t), \quad t \ge -s, \\ X(-s) &= x. \end{cases} \tag{3.48}$$

Here $\overline{W}(t)$ is defined for all $t \in \mathbb{R}$ as follows. We take another cylindrical process $W_1(t)$ independent of $W(t)$, we set

$$\overline{W}(t) = \begin{cases} W(t) \text{ if } t \ge 0, \\ W_1(-t) \text{ if } t \le 0, \end{cases}$$

and we denote by $\overline{\mathscr{F}}_t$ the $\sigma$-algebra generated by $\overline{W}(s), s \le t$, $t \in \mathbb{R}, k \in \mathbb{N}$.

Now Theorem 3.5 can be generalized in a straightforward way to solve problem (3.48). We denote by $X(t, -s, x)$ its unique mild solution, that is the solution of the integral equation

$$X(t,-s,x) = e^{(t+s)A}x + \int_{-s}^{t} e^{(t-u)A}F(X(u,-s,x))du + W_A(t,-s),$$

where

$$W_A(t,-s) = \int_{-s}^{t} e^{(t-u)A}Bd W(u).$$

Obviously, the law of $X(0, -t, x)$ coincides with that of $X(t, x)$.

We are going to show, following [50], that the law $\mathscr{L}(X(t,x))$ is convergent as $t \to +\infty$ to the unique invariant measure $\nu$ of $P_t$. In fact we shall prove more, namely that there exists the limit

$$\lim_{s\to+\infty} X(0,-s,x) := \eta \quad \text{in } L^2(\Omega, \mathscr{F}, \mathbb{P}; H).$$

The law of $\eta$ will be the required invariant measure for $P_t$.

**Proposition 3.16.** *Assume that Hypothesis 3.14 holds. Then there exists* $\zeta \in L^2(\Omega, \mathscr{F}, \mathbb{P}; H)$ *such that for any* $x \in H$ *we have*

$$\lim_{s \to +\infty} X(0, -s, x) = \zeta \quad \text{in } L^2(\Omega, \mathscr{F}, \mathbb{P}; H). \tag{3.49}$$

*Moreover, there exists* $c_1 > 0$ *such that*

$$\mathbb{E}(|X(0, -s, x) - \zeta|^2) \le c_1 e^{-2\omega_1 s}, \quad s > 0. \tag{3.50}$$

*Proof.* Let $x \in H$ be fixed and set $X_s(t) = X(t, -s, x)$ for all $t \ge s$. Then we have

$$X_s(t) = e^{(t+s)A}x + \int_{-s}^{t} e^{(t-u)A} F(X_s(u))du + W_A(-s, t), \quad t \ge s.$$

Setting $Y_s(t) = X_s(t) - W_A(t, -s)$, we find that

$$Y_s(t) = e^{(t+s)A}x + \int_{-s}^{t} e^{(t-u)A} F(Y_s(u) + W_A(u, -s))du.$$

Therefore $Y_s(t)$ is the mild solution of the initial value problem

$$\begin{cases} \dfrac{d}{dt} Y_s(t) = AY_s(t) + F(Y_s(t) + W_A(t, -s)), \quad t \ge -s, \\[2mm] Y(-s) = x. \end{cases} \tag{3.51}$$

We divide the remainder of the proof in four steps.

*Step* 1. There exists $C_1 > 0$ such that

$$\mathbb{E}(|X(t, -s, x)|^2) \le C_1(1 + e^{-\omega_1(t+s)}|x|^2), \quad t \ge s.$$

Multiplying scalarly both sides of the first identity in (3.51) by $Y_s(t)$, we obtain

$$\frac{1}{2} \frac{d}{dt} |Y_s(t)|^2 = \langle AY_s(t) + F(Y_s(t) + W_A(t, -s)) - F(W_A(t, -s)), Y_s(t) \rangle$$

$$+ \langle F(W_A(t, -s)), Y_s(t) \rangle.$$

Since

$$\langle Ay + F(y + W_A(t, -s)) - F(W_A(t, -s)), y \rangle \le -\omega_1 |y|^2, \quad y \in D(A),$$

we deduce that

$$\frac{1}{2} \frac{d}{dt} |Y_s(t)|^2 \le -\omega_1 |Y_s(t)|^2 + \langle F(W_A(t, -s)), Y_s(t) \rangle$$

$$\le -\frac{\omega_1}{2} |Y_s(t)|^2 + \frac{2}{\omega_1} |F(W_A(t, -s))|^2.$$

By the usual comparison result it follows that

$$|Y_s(t)|^2 \le e^{-\omega_1(t+s)}|x|^2 + \frac{4}{\omega_1} \int_{-s}^t e^{-\omega_1(t-u)}|F(W_A(u,-s))|^2 du$$

and consequently, since $|F(x)|^2 \le 2|F(0)|^2 + 2K^2|x|^2$, we obtain

$$|X(t,-s,x)|^2 \le 2e^{-\omega_1(t+s)}|x|^2 + \frac{16K^2}{\omega_1} \int_{-s}^t e^{-\omega_1(t-u)}|W_A(u,-s)|^2 du$$

$$+ 2|W_A(t,-s)|^2 + \frac{16|F(0)|^2}{\omega_1} \int_{-s}^t e^{-\omega_1(t-u)} du.$$

$$(3.52)$$

On the other hand we have

$$[\mathbb{E}|W_A(u,-s)|]^2 = \int_{-s}^t \text{Tr}\, [e^{(t-u)A}Ce^{(t-u)A^*}]du$$

$$(3.53)$$

$$= \int_0^{t+s} \text{Tr}\, [e^{vA}Ce^{vA^*}]dv \le \text{Tr}\, Q_\infty.$$

Now, the conclusion follows from (3.52) and (3.53).

*Step 2.* There exists $\zeta_x \in L^2(\Omega, \mathscr{F}, \mathbb{P}; H)$ such that $\lim_{s \to +\infty} X(0,-s,x) = \zeta_x$ in $L^2(\Omega, \mathscr{F}, \mathbb{P}; H)$.

Let $s > s_1$, and set $Z(t) = X_s(t) - X_{s_1}(t)$. Then we have

$$\begin{cases} \dfrac{d}{dt} Z(t) = AZ(t) + F(X_s(t)) - F(X_{s_1}(t)), & t \ge -s_1, \\[2mm] Z(-s_1) = X_s(-s_1) - x. \end{cases}$$

Multiplying scalarly both sides of the first equation by $Z(t)$ we obtain

$$\frac{1}{2} \frac{d}{dt} |Z(t)|^2 \le -\omega_1|Z(t)|^2, \quad t \ge 0,$$

so that

$$|X_s(t) - X_{s_1}(t)|^2 = |Z(t)|^2 \le e^{-2\omega_1(t+s)}|X_s(-s_1) - x|^2.$$

Recalling Step 1, we see that there exists a constant $c_1 > 0$ such that

$$\mathbb{E}\left(|X_s(0) - X_{s_1}(0)|^2\right) \le C_1 e^{-2\omega_1 s}, \quad s > s_1. \qquad (3.54)$$

Consequently $\{X_s(0)\}$ is a Cauchy sequence in $L^2(\Omega, \mathscr{F}, \mathbb{P}; H)$ and Step 2 is proved.

*Step 3.* $\zeta_x$ does not depend of $x$.

Let $x, y \in H$, and set

$$\rho_s(t) = X(t, -s, x) - X(t, -s, y).$$

Then we have

$$\begin{cases} \dfrac{d}{dt}\,\rho_s(t) = A\rho_s(t) + F(X(t, -s, x)) - F(X(t, -s, y)), \\[2mm] \rho(-s) = x - y. \end{cases}$$

Multiplying scalarly the first equation by $\rho_s(t)$ and arguing as before, it follows that

$$|\rho_s(t)| \le e^{-\omega_1(t+s)}|x - y|, \quad t \ge -s.$$

Therefore, as $s \to +\infty$ we obtain $\zeta_x = \zeta_y$ as required.

*Step 4.* (3.50) holds.

This follows from (3.54). $\qquad\qquad\qquad\qquad\qquad\qquad\qquad\qquad\qquad\quad\;\Box$

We can prove now the following result.

**Theorem 3.17.** *Assume that Hypothesi 3.14 holds and let $\nu$ be the law of the random variable $\zeta$ defined by (3.49). Then the following statements hold:*

(i) *We have*

$$\lim_{t \to +\infty} P_t\varphi(x) = \int_H \varphi(y)\nu(dy), \quad x \in H, \; \varphi \in C_b(H). \qquad (3.55)$$

(ii) *$\nu$ is the unique invariant measure for $P_t$.*

(iii) *For any Borel probability measure $\lambda$ on $H$ we have*

$$\lim_{t \to +\infty} \int_H P_t\varphi(x)\lambda(dx) = \int_H \varphi(x)\nu(dx), \quad \varphi \in C_b(H).$$

(iv) *There exists $c > 0$ such that for any function $\varphi \in C_b^1(H)$ we have*

$$\left| P_t\varphi(x) - \int_H \varphi(y)\nu(dy) \right| \le c\|\varphi\|_1 e^{-\omega_1 t}, \quad t \ge 0. \qquad (3.56)$$

*Proof.* (i) If $\varphi \in C_b(H)$ we have

$$P_t\varphi(x) = \mathbb{E}[\varphi(X(t, x)] = \mathbb{E}[\varphi(X(0, -t, x))].$$

Letting $t$ tend to $+\infty$ yields

$$\lim_{t \to +\infty} P_t \varphi(x) = \mathbb{E}[\varphi(\zeta)] = \int_H \varphi(y)\nu(dy),$$

since the law of $\zeta$ is precisely $\nu$.

(ii) For any $t, s > 0$ and any $\varphi \in C_b(H)$, we have

$$\int_H P_{t+s}\varphi(x)\nu(dx) = \int_H P_t P_s \varphi(x)\nu(dx).$$

Letting $t$ tend to $+\infty$ we find, by (3.55),

$$\int_H \varphi(y)\nu(dy) = \int_H P_s \varphi(y)\nu(dy),$$

so that $\nu$ is invariant.

Let us prove uniqueness of $\nu$. Let $\lambda$ be an invariant probability measure for $P_t$, that is

$$\int_H P_t \varphi(x)\lambda(dx) = \int_H \varphi(x)\lambda(dx), \quad \varphi \in C_b(H).$$

Then, letting $t$ tend to $+\infty$ yields, thanks to (3.55),

$$\int_H \varphi(x)\nu(dx) = \int_H \varphi(x)\lambda(dx), \quad \varphi \in C_b(H),$$

which implies $\lambda = \nu$.

(iii) follows again by (3.55). Let us prove finally (iv). Since

$$P_t \varphi(x) - \int_H \varphi(y)\nu(dy) = \mathbb{E}(\varphi(X(0, -t, x)) - \varphi(\zeta)),$$

then, taking into account (3.50), we have

$$\left| P_t \varphi(x) - \int_H \varphi(y)\nu(dy) \right| \le \|\varphi\|_1 \mathbb{E}|\varphi(X(0, -t, x)) - \varphi(\zeta)| \le \|\varphi\|_1 \sqrt{c_1} e^{-\omega_1 t},$$

$t \ge 0$. □

Let us prove now that all moments of $\nu$ are finite.

**Proposition 3.18.** *Assume that Hypothesis 3.14 holds. Then for any $n \in \mathbb{N}$ we have*

$$\int_H |x|^{2n}\nu(dx) < +\infty. \tag{3.57}$$

*Proof.* By (3.52) with $s = 0$ we have

$$|X(t,x)|^2 \leq 2e^{-\omega_1 t}|x|^2 + \frac{16K^2}{\omega_1} \int_0^t e^{-\omega_1(t-u)}|W_A(u)|^2 du$$

$$+2|W_A(t)|^2 + \frac{16|F(0)|^2}{\omega_1} \int_0^t e^{-\omega_1 u} du.$$

Taking into account that $W_A(t)$ is a Gaussian random variable $N_{Q_t}$, we see that there exists $c_n > 0$ such that

$$\mathbb{E}\left(|X(t,x)|^{2n}\right) \leq c_n(1 + e^{-\omega_1 t}|x|^{2n}), \quad x \in H, \ t \geq 0. \tag{3.58}$$

Denote by $\nu_{t,x}$ the law of $X(t,x)$. Then by (3.58) it follows that for any $\beta > 0$,

$$\int_H \frac{|y|^{2n}}{1 + \beta|y|^{2n}} \nu_{t,x}(dy) \leq \int_H |y|^{2n} \nu_{t,x}(dy)$$

$$= \mathbb{E}(|X(t,x)|^{2n}) \leq c_n(1 + e^{-n\omega_1 t}|x|^{2n}), \quad x \in H, \ t \geq 0.$$

Consequently, letting $t$ tend to $\infty$ and taking into account that $\nu_{t,x}$ weakly converges to $\nu$ by Proposition 3.16, we find that

$$\int_H \frac{|y|^{2n}}{1 + \beta|y|^{2n}} \nu(dy) \leq c_n,$$

which, letting $\beta$ tend to 0, yields (3.57). $\qquad \square$

## 3.5 The transition semigroup in $L^2(H, \nu)$

In this section we still assume that Hypothesis 3.14 holds. We denote by $\nu$ the unique invariant measure for the transition semigroup $P_t$.

Proceeding as for the proof of Proposition 2.36, we see that $P_t$ can be uniquely extended to a strongly continuous semigroup of contractions in $L^p(H, \nu)$, $p \geq 1$, whose infinitesimal generator we denote by $K_p$.

The main result of this section is that $K_2$ is the closure of the Kolmogorov operator $K_0$ defined by

$$K_0\varphi = \frac{1}{2} \text{Tr} [CD^2\varphi] + \langle x, A^*D\varphi \rangle + \langle F(x), D\varphi \rangle = L\varphi + \langle F(x), D\varphi \rangle, \ \varphi \in \mathscr{E}_A(H),$$

or, in other words, that $\mathscr{E}_A(H)$ is a core for $K_2$.

We need the following result.

**Proposition 3.19.** *For any* $\varphi \in \mathscr{E}_A(H)$ *we have that*

$$\mathbb{E}\left[\varphi(X(t,x))\right] = \varphi(x) + \mathbb{E}\left[\int_0^t K_0\varphi(X(s,x))ds\right], \quad t \geq 0, \ x \in H. \tag{3.59}$$

Notice that this formula is meaningful since $K_0\varphi$ has linear growth on $x$. It follows from Itô's formula (see [49]) but we prefer to give here a direct, more elementary proof.

*Proof of Proposition* 3.19. Fix $t \geq 0, x \in H$ and let $h > 0$. Set

$$X(t+h, x) - X(t, x) = \delta_h.$$

Then by the Taylor formula we have

$$\frac{1}{h} \left[ \mathbb{E}\varphi(X(t+h, x)) - \mathbb{E}\varphi(X(t, x)) \right] = I_1 + I_2 + I_3,$$

where

$$I_1 = \frac{1}{h} \, \mathbb{E}\langle D\varphi(X(t, x)), \delta_h \rangle,$$

$$I_2 = \frac{1}{2h} \, \mathbb{E}\langle D^2\varphi(X(t, x)) \cdot \delta_h, \delta_h \rangle,$$

$$I_3 = \frac{1}{h} \, \mathbb{E} \int_0^1 (1-\xi)\langle [D^2\varphi((1-\xi)X(t, x) + \xi X(t+h, x)) - D^2\varphi(X(t, x))]\delta_h, \delta_h \rangle d\xi.$$

On the other hand we have

$$\delta_h = (e^{hA} - 1)X(t, x) + \int_t^{t+h} e^{(t+h-s)A} F(X(s, x))ds + W_A(t+h, t),$$

where

$$W_A(t+h, t) = \int_t^{t+h} e^{(t+h-s)A} B \, dW(s).$$

Therefore, taking into account that $W_A(t+h, t)$ is independent of $D\varphi(X(t, x))$, we find that

$$
\begin{aligned}
I_1 &= \mathbb{E}\left\langle D\varphi(X(t, x)), \frac{1}{h} \int_t^{t+h} e^{(t+h-s)A} F(X(s, x))ds \right\rangle \\
&\quad + \mathbb{E}\left\langle \frac{1}{h}(e^{hA^*} - 1)D\varphi(X(t, x)), X(t, x) \right\rangle.
\end{aligned}
$$

Thus, letting $h$ tend to 0, we find that

$$\lim_{h \to 0} I_1 = \mathbb{E}\left[\langle X(t, x), A^* D\varphi(X(t, x))\rangle\right] + \mathbb{E}\left[\langle D\varphi(X(t, x)), F(X(t, x))\rangle\right]. \quad (3.60)$$

Concerning $I_2$ we have

$$I_2 = \frac{1}{2} \, \mathbb{E}\left\langle D^2\varphi(X(t,x)) \cdot \left[\int_t^{t+h} e^{(t+h-s)A} F(X(s,x))ds\right], \right.$$

$$\left. \frac{1}{h}\int_t^{t+h} e^{(t+h-s)A} F(X(s,x))ds \right\rangle$$

$$+\mathbb{E}\left\langle D^2\varphi(X(t,x)) \cdot \left[\frac{1}{h}\int_t^{t+h} e^{(t+h-s)A} F(X(s,x))ds\right], W_A(t+h,t) \right\rangle$$

$$+\frac{1}{2h} \, \mathbb{E}\langle D^2\varphi(X(t,x)) \cdot W_A(t+h,t), W_A(t+h,t)\rangle$$

$$:= I_{2,1} + I_{2,2} + I_{2,3}.$$

Clearly

$$\lim_{h\to 0} I_{2,1} = \lim_{h\to 0} I_{2,2} = 0, \tag{3.61}$$

and

$$I_{2,3} = \frac{1}{2h} \, \mathbb{E}\langle D^2\varphi(X(t,x)) \cdot W_A(t+h,t), W_A(t+h,t)\rangle$$

$$= \frac{1}{2h} \sum_{h,k=1}^{\infty} \mathbb{E}[D_h D_k \varphi(X(t,x))]\mathbb{E}[\langle W_A(t+h,t), e_h\rangle\langle W_A(t+h,t), e_k\rangle]$$

$$= \frac{1}{2h} \sum_{h,k=1}^{\infty} \mathbb{E}[D_h D_k \varphi(X(t,x))]\langle Q_{t+h,t} e_h, e_k\rangle$$

$$= \frac{1}{2h} \, \mathbb{E}[\mathrm{Tr}\,(Q_{t+h,t} D^2\varphi(X(t,x)))].$$

Therefore

$$\lim_{h\to 0} I_{2,3} = \frac{1}{2} \, \mathbb{E}[\mathrm{Tr}\,(Q_\infty D^2\varphi(X(t,x)))]. \tag{3.62}$$

Finally, after some standard manipulations, we find that

$$\lim_{h\to 0} I_3 = 0. \tag{3.63}$$

Now the conclusion follows from (3.61)–(3.63). $\qquad\square$

For the next result we shall need the following Lumer–Phillips Theorem, see [55], concerning dissipative operators. We recall that a linear operator $\mathscr{A} : D(\mathscr{A}) \subset \mathscr{H} \to \mathscr{H}$ in a Hilbert space $\mathscr{H}$ is *dissipative* if

$$|\varphi|_{\mathscr{H}} \leq \frac{1}{\lambda} |\lambda\varphi - \mathscr{A}\varphi|_{\mathscr{H}}, \quad \forall \varphi \in D(\mathscr{A}), \ \lambda > 0.$$

Any dissipative operator is closable. The dissipative operator $\mathscr{H}$ is called *m-dissipative* if the range of $\lambda - \mathscr{A}$ coincides with $\mathscr{H}$ for some (and consequently for any) $\lambda > 0$. An operator $\mathscr{A}$ with dense domain is *m*-dissipative if and only if it is the infinitesimal generator of a strongly continuous semigroup of contractions in $\mathscr{H}$.

**Theorem 3.20 (Lumer–Phillips).** *Let $\mathscr{A} : D(\mathscr{A}) \subset \mathscr{H} \to \mathscr{H}$ be a dissipative operator in the Hilbert space in $\mathscr{H}$ such that $D(\mathscr{A})$ is dense in $\mathscr{H}$. Assume that for some $\lambda > 0$ the range of $\lambda - \mathscr{A}$ is dense in $\mathscr{H}$. Then the closure of $\mathscr{A}$ is m-dissipative.*

We can now prove

**Theorem 3.21.** *Assume that Hypothesis 3.14 holds. Then $K_2$ is the closure of $K_0$ in $L^2(H, \nu)$.*

*Proof.* We start by proving that $K_2\varphi = K_0\varphi$ for all $\varphi \in \mathscr{E}_A(H)$. In fact, by Proposition 3.19 it follows that for all $\varphi \in \mathscr{E}_A(H)$ we have

$$K_2\varphi(x) = \lim_{t \to 0} \frac{1}{t} \left( P_t\varphi(x) - \varphi(x) \right) = K_0\varphi(x), \quad x \in H.$$

By Proposition 3.18, $K_2\varphi \in L^2(H, \nu)$, thus, by (3.59) and the dominated convergence theorem we have

$$K_2\varphi = \lim_{t \to 0} \frac{1}{t} \left( P_t\varphi - \varphi \right) = K_0\varphi, \quad \text{in } L^2(H, \nu).$$

Therefore, $K_2$ extends $K_0$. Since $K_2$ is dissipative, so is $K_0$. Consequently, $K_0$ is closable. Let us denote by $\overline{K_0}$ its closure. It remains to show that $K_2 = \overline{K_0}$.

Let $\lambda > 0$, $f \in \mathscr{E}_A(H)$ and set

$$\varphi(x) = \int_0^{+\infty} e^{-\lambda t} P_t f(x) dt, \quad x \in H.$$

We are going to show that $\varphi$ belongs to the the range of $\lambda - \overline{K_0}$. For this we proceed in several steps.

*Step* 1. Assume that $F \in C_b^3(H; H)$. Then $\varphi \in D(L, C_{b,2}(H)) \cap C_b^1(H)$ and we have [4]

$$\lambda\varphi - L\varphi - \langle F(x), D\varphi \rangle = f.$$

---

[4] Recall that $D(L, C_{b,2}(H))$ was defined in §2.8.3

We first notice that $\varphi \in C_b^2(H)$ by Theorem 3.7 and that

$$D\varphi(x) = \int_0^{+\infty} e^{-\lambda t} \mathbb{E}\left[ (DX(t,x))^* Df(X(t,x)) \right] dt, \quad x \in H.$$

To prove that $\varphi \in D(L, C_{b,2}(H))$ we have to show the existence of the derivative $\frac{d}{dh} R_h\varphi(x)$ at $h = 0$, where

$$R_h\varphi(x) = \mathbb{E}[\varphi(Z(h,x))],$$

and

$$Z(h,x) = e^{hA}x + W_A(h), \quad x \in H, \ h > 0$$

and that

$$\sup_{h\in(0,1]} \frac{d}{dh} \|R_h\varphi - \varphi\|_{0,2} < +\infty.$$

We have in fact

$$R_h\varphi(x) = \mathbb{E}[\varphi(Z(h,x))] = \mathbb{E}\left[ \varphi\left( X(h,x) - \int_0^h e^{(h-s)A} F(X(s,x))ds \right) \right]$$

$$= P_h\varphi(x) - \left\langle D\varphi(X(h,x)), \int_0^h e^{(h-s)A} F(X(s,x))ds \right\rangle + o(h),$$

where $\lim_{h\to 0} \frac{o(h)}{h} = 0$. It follows that

$$R_h\varphi(x) = e^{\lambda h} \int_h^{+\infty} e^{-\lambda t} P_t f(x)dt$$

$$- \left\langle \int_0^{+\infty} e^{-\lambda t} DP_t f(x)dt, \int_0^h e^{(h-s)A} F(X(s,x))]ds \right\rangle + o(h).$$

Therefore, taking the derivative at $h = 0$, we see that $\varphi \in D(L, C_{b,2}(H))$ and

$$L\varphi(x) = \lambda\varphi(x) - f(x) - \langle D\varphi(x), F(x) \rangle.$$

Notice that $L\varphi \in C_{b,2}^1(H)$.

*Step 2.* $\varphi \in D(\overline{K_0})$.

Since we have proved that $\varphi \in D(L, C_{b,2}(H)) \cap C_{b,2}^1(H)$ and $L\varphi \in C_{b,2}^1(H)$, then by Proposition 2.68 there exists a three-index sequence $\{\varphi_{n_1,n_2,n_3}\} \subset \mathscr{E}_A(H)$ such that

$$\lim_{n_1\to\infty} \lim_{n_2\to\infty} \lim_{n_3\to\infty} \varphi_{n_1,n_2,n_3}(x) = \varphi(x), \quad x \in H,$$

$$\lim_{n_1\to\infty} \lim_{n_2\to\infty} \lim_{n_3\to\infty} D\varphi_{n_1,n_2,n_3}(x) = D\varphi(x), \quad x \in H,$$

$$\lim_{n_1\to\infty}\lim_{n_2\to\infty}\lim_{n_3\to\infty} L\varphi_{n_1,n_2,n_3}(x) = L\varphi(x), \quad x \in H.$$

By the dominated convergence theorem it follows that

$$\lim_{n_1\to\infty}\lim_{n_2\to\infty}\lim_{n_3\to\infty} K_0\varphi_{n_1,n_2,n_3}(x) = \overline{K_0}\varphi(x) = L\varphi + \langle F(x), D\varphi\rangle, \quad x \in H.$$

Consequently we have

$$\lambda\varphi - \overline{K_0}\varphi = f.$$

*Step* 3. Conclusion.

Let us introduce a regularization of $F$ as in (3.28),

$$\langle F_\beta(x), h\rangle = \int_H \langle F(e^{\beta S}x + y), e^{\beta S}h\rangle N_{\frac{1}{2} S^{-1}(e^{2\beta S}-1)}(dy), \quad \beta > 0$$

and consider the equation

$$\lambda\varphi_\beta - L\varphi_\beta - \langle F_\beta(x), D\varphi_\beta\rangle = f.$$

We know by Step 2 that $\varphi_\beta \in D(\overline{K_0})$ and

$$\lambda\varphi_\beta - \overline{K_0}\varphi_\beta = f + \langle F_\beta - F, D\varphi_\beta\rangle.$$

We claim that

$$\lim_{\beta\to0}\langle F_\beta - F, D\varphi_\beta\rangle = 0 \quad \text{in } L^2(H,\nu). \tag{3.64}$$

We have in fact

$$\|D\varphi_\beta\|_0 \le \frac{1}{\lambda + \omega_1}\|Df\|_0,$$

and, since $|F(x)| \le K|x| + |F(0)|$,

$$|F_\beta(x)| \le \int_H [K|e^{\beta S}x + y| + |F(0)|]N_{\frac{1}{2} S^{-1}(e^{2\beta S}-1)}(dy) \le a + b|x|,$$

for suitable constants $a, b$. Now the dominated convergence theorem implies (3.64).

In conclusion, we have proved that the closure of the range of $\overline{K_0}$ includes $\mathscr{E}_A(H)$, so that it is dense in $L^2(H,\nu)$ and, by the Lumer–Phillips theorem Theorem 3.20, $\overline{K_0}$ is $m$-dissipative. Since $K_2$ is $m$-dissipative as well (it is the infinitesimal generator of a strongly continuous semigroup of contractions) and extends $\overline{K_0}$, it must coincide with $\overline{K_0}$ as claimed.     □

## 3.6   The integration by parts formula and its consequences

In this section we still assume that Hypothesis 3.14 holds, we set $\omega_1 := -\omega - \kappa$, we denote by $\nu$ the invariant measure of the transition semigroup $P_t$ and by $K_2$ its infinitesimal generator in $L^2(H,\nu)$.

Since the proofs of the two following propositions are completely similar to that of Propositions 2.39 and 2.43 respectively, they will be omitted.

**Proposition 3.22.** *The operator*

$$D_C : \mathscr{E}_A(H) \to C_b(H; H), \quad \varphi \to C^{1/2} D\varphi,$$

*is uniquely extendible to a linear bounded operator* $D_C \colon D(K_2) \to L^2(H, \nu; H)$. *Moreover, the following identity holds:*

$$\int_H K_2\varphi \; \varphi \; d\nu = -\frac{1}{2} \int_H |D_C\varphi|^2 d\nu, \quad \varphi \in D(K_2). \tag{3.65}$$

**Proposition 3.23.** *Let* $\varphi \in L^2(H, \nu)$ *and* $t \geq 0$. *Then, for any* $T > 0$, *the linear operator*

$$D_C : D(K_2) \to L^2(0, T; L^2(H, \nu; H)), \quad \varphi \to D_C P_t\varphi,$$

*is uniquely extendible to a linear bounded operator, still denoted by* $D_C$, *from* $L^2(H, \nu)$ *into* $L^2(0, T; L^2(H, \nu; H))$. *Moreover the following identity holds:*

$$\int_H (P_t\varphi)^2 \; d\nu + \int_0^t ds \int_H |D_C P_s\varphi|^2 d\nu = \int_H \varphi^2 \; d\nu. \tag{3.66}$$

## 3.6.1 The Sobolev space $W^{1,2}(H, \nu)$

Here we assume in addition that $C = I$ and $F \in C_b^2(H; H)$, but all results we are going to prove hold if $C^{-1} \in L(H)$. We want to show that the mapping

$$D : \mathscr{E}_A(H) \subset L^2(H, \nu) \to L^2(H, \nu; H), \quad \varphi \to D\varphi$$

is closable. For this we need a lemma, see [37].

**Lemma 3.24.** *Let* $\{\varphi_n\} \subset \mathscr{E}_A(H)$ *and* $G \in L^2(H, \nu; H)$ *be such that*

$$\lim_{n \to \infty} D\varphi_n = G \quad \text{in } L^2(H, \nu; H).$$

*Then, for any* $t \geq 0$ *we have*

$$\lim_{n \to \infty} DP_t\varphi_n = \mathbb{E}[X_x^*(t, x)G(X(t, x))] \quad \text{in } L^2(H, \nu; H).$$

*In particular, if* $D\varphi_n \to 0$ *in* $L^2(H, \nu; H)$ *we have* $DP_t\varphi_n \to 0$ *in* $L^2(H, \nu; H)$ *for all* $t > 0$.

*Proof.* We have in fact, by Theorem 3.6,

$$DP_t\varphi_n(x) = E[X_x^*(t, x)D\varphi_n(X(t, x))], \quad t \geq 0, \; x \in H,$$

so that, taking into account the invariance of $\nu$,

$$\int_H |DP_t\varphi_n(x) - \mathbb{E}[X_x^*(t,x)G(X(t,x))]|^2 \, \nu(dx)$$

$$= \int_H |\mathbb{E}[X_x^*(t,x)(D\varphi_n(X(t,x)) - G(X(t,x)))]|^2 \, \nu(dx)$$

$$\leq e^{-2\omega_1 t} \int_H \mathbb{E}[|D\varphi_n(X(t,x)) - G(X(t,x))|^2] \nu(dx)$$

$$= e^{-2\omega_1 t} \int_H P_t(|D\varphi_n - G|^2)(x)\nu(dx)$$

$$= e^{-2\omega_1 t} \int_H |D\varphi_n(x) - G(x)|^2 \nu(dx).$$

The conclusion of the lemma follows.    $\square$

**Theorem 3.25.** *D is closable. Moreover, if $\varphi$ belongs to the domain $\overline{D}$ of the closure of $D$ and $\overline{D}\varphi = 0$, we have that $\overline{D}P_t\varphi = 0$ for any $t > 0$.*

*Proof.* Let $\{\varphi_n\} \subset \mathscr{E}_A(H)$ and $G \in L^2(H,\nu;H)$ be such that

$$\varphi_n \to 0 \quad \text{in } L^2(H,\nu), \quad D\varphi_n \to G \quad \text{in } L^2(H,\nu;H).$$

By (3.66) we have that

$$\int_H (P_t\varphi_n)^2 \, d\nu + \int_0^t ds \int_H |DP_s\varphi_n|^2 d\nu = \int_H \varphi_n^2 \, d\nu.$$

Letting $n \to \infty$ yields

$$\lim_{n\to\infty} \int_0^t ds \int_H |DP_s\varphi_n|^2 d\nu = 0.$$

Consequently, by Lemma 3.24, it follows that

$$\int_0^t ds \int_H (\mathbb{E}[X_x^*(s,x)G(X(s,x))])^2 \, \nu(dx) = 0, \quad h \in H.$$

Then for almost all $t \geq 0$ we have that

$$\mathbb{E}[X_x^*(s,x)G(X(s,x))] = 0. \tag{3.67}$$

Now fix $h \in H$. Then we have

$$|\mathbb{E}[\langle G(X(t,x)), h\rangle]|$$

$$\leq |\mathbb{E}[\langle G(X(t,x)), X_x(t,x) \cdot h\rangle]| + |\mathbb{E}[\langle G(X(t,x)), h - X_x(t,x) \cdot h\rangle]|$$

$$= |\mathbb{E}[\langle G(X(t,x)), h - X_x(t,x) \cdot h\rangle]|.$$

Taking into account the invariance of $\nu$ and (3.67), we find that

$$\int_H |P_t(\langle G(x), h\rangle)|\nu(dx) = \int_H |\mathbb{E}[\langle G(X(t,x)), h\rangle]|\nu(dx)$$

$$= \int_H |\mathbb{E}[\langle G(X(t,x)), h - X_x(t,x) \cdot h\rangle]|\nu(dx)$$

$$\leq \left[\int_H \mathbb{E}[|G(X(t,x))|^2]\nu(dx)\right]^{1/2} \left[\int_H \mathbb{E}[|h - X_x(t,x) \cdot h|^2]\nu(dx)\right]^{1/2}.$$

Therefore, as $t \to 0$ we find by the strong continuity of $P_t$ in $L^1(H, \nu)$,

$$\int_H |\langle G(x), h\rangle|\nu(dx) = 0,$$

and by the arbitrariness of $h$ it follows that $G = 0$ as required. Finally, the last statement follows from Lemma 3.24. $\qquad\square$

By Theorem 3.25 it follows that the mapping

$$D : \mathscr{E}_A(H) \subset L^2(H, \nu) \to L^2(H, \nu; H), \quad \varphi \mapsto D\varphi$$

is closable, let $\overline{D}$ be its closure. We shall denote by $W^{1,2}(H, \nu)$ the domain of $\overline{D}$ and, if there is no possibility of confusion, we shall set $\overline{D} = D$.

**Proposition 3.26.** *We have $D(K_2) \subset W^{1,2}(H, \nu)$ with continuous embedding. Moreover, the following identity holds:*

$$\int_H K_2\varphi \, \varphi \, d\nu = -\frac{1}{2}\int_H |D\varphi|^2 d\nu, \quad \varphi \in D(K_2). \tag{3.68}$$

*Proof.* Let $\varphi \in D(K_2)$. Since $\mathscr{E}_A(H)$ is a core for $K_2$, there exists a sequence $\{\varphi_n\} \subset \mathscr{E}_A(H)$ such that

$$\varphi_n \to \varphi, \quad K_0\varphi_n \to K_2\varphi \quad \text{in } L^2(H, \nu).$$

By (3.65) it follows that

$$\int_H |D(\varphi_n - \varphi_m)|^2 d\nu \leq 2\int_H |K_0(\varphi_n - \varphi_m)| \, |\varphi_n - \varphi_m| \, d\nu.$$

Therefore the sequence $\{D\varphi_n\}$ is Cauchy in $L^2(H, \nu; H)$. Since $D$ is closed it follows that $\varphi \in W^{1,2}(H, \nu)$ as required. $\qquad\square$

### 3.6.2  Poincaré and log-Sobolev inequalities, spectral gap

We use here the notation

$$\overline{\varphi} = \int_H \varphi d\nu.$$

To prove the Poincaré inequality we need a lemma.

**Lemma 3.27.** *For any $\varphi \in L^2(H, \nu)$ we have*

$$\int_H |\varphi - \overline{\varphi}|^2 d\nu = \int_0^{+\infty} ds \int_H |DP_s\varphi|^2 d\nu. \tag{3.69}$$

*Moreover, if $\varphi \in W^{1,2}(H, \nu)$ is such that $\overline{D}\varphi = 0$, we have $\varphi = \overline{\varphi}$.*

*Proof.* Equation (3.69) follows letting $t \to +\infty$ in (3.66). Now, let $\varphi \in W^{1,2}(H, \nu)$ be such that $\overline{D}\varphi = 0$. Then, by Theorem 3.25 we have that $\overline{D}P_s\varphi = 0$ for all $s \in [0, t]$. Consequently, by (3.69) we find $\varphi = \overline{\varphi}$.   □

Let us prove now the Poincaré inequality.

**Proposition 3.28.** *For any $\varphi \in W^{1,2}(H, \nu)$ we have*

$$\int_H |\varphi - \overline{\varphi}|^2 d\nu \le \frac{1}{2\omega_1} \int_H |D\varphi|^2 d\nu. \tag{3.70}$$

*Proof.* Let first $\varphi \in \mathscr{E}_A(H)$. Then for any $h \in H$ and $s \ge 0$ we have

$$\langle DP_s\varphi(x), h \rangle = \mathbb{E}\left[\langle D\varphi(X(s, x)), h \rangle\right].$$

Taking into account (3.47) it follows that

$$\mathbb{E}[|\langle DP_s\varphi(x), h \rangle|^2] \le \mathbb{E}\left[|D\varphi(X(s, x))|^2 |\eta^h(s, x)|^2\right]$$

$$\le e^{-2\omega_1 s}\mathbb{E}\left[|D\varphi(X(s, x))|^2\right] |h|^2 = e^{-2\omega_1 s}P_s(|D\varphi|^2)(x) |h|^2.$$

The arbitrariness of $h$ yields

$$|DP_s\varphi(x)|^2 \le e^{-2\omega_1 s}P_s(|D\varphi|^2)(x), \quad x \in H \; s \ge 0.$$

From (3.69), taking into account the invariance of $\nu$, we obtain

$$\int_H |\varphi(x) - \overline{\varphi}|^2 \nu(dx) \le \int_0^{+\infty} e^{-2\omega_1 s}ds \int_H P_s(|D\varphi|^2)(x)\nu(dx)$$

$$\le \int_0^{+\infty} e^{-2\omega_1 s}ds \int_H |D\varphi(x)|^2 \nu(dx),$$

and the conclusion follows. If $\varphi \in W^{1,2}(H, \nu)$, we proceed by density.   □

Now we can show that the spectrum $\sigma(K_2)$ of $K_2$ consists of $0$ and a set included in the half–space

$$\{\lambda \in \mathbb{C} : \Re\, \lambda \le -\omega_1\}$$

(*spectral gap*). The spectral gap implies an exponential convergence of $P_t\varphi$ to the equilibrium

$$\int_{\mathbb{R}} |P_t\varphi - \overline{\varphi}|^2 d\nu \le e^{-2\omega_1 t} \int_{\mathbb{R}} |\varphi|^2 d\nu, \quad \varphi \in L^2(\mathbb{R}, \nu). \tag{3.71}$$

The following result holds in fact:

**Proposition 3.29.** *We have*

$$\sigma(K_2)\backslash\{0\} \subset \{\lambda \in \mathbb{C} : \Re\, \lambda \le -\omega_1\}, \tag{3.72}$$

*and (3.71) is fulfilled.*

*Proof.* Let us consider the subspace of $L^2(H, \nu)$,

$$L_0^2(H, \nu) = \left\{\varphi \in L^2(H, \nu) : \overline{\varphi} = 0\right\}.$$

Clearly $L_0^2(H, \nu)$ is an invariant subspace for $P_t$. Moreover, by (3.70),

$$\langle K_2\varphi, \varphi\rangle_{L^2 H, \nu)} = -\frac{1}{2}\int_H |D\varphi|^2 d\nu \le -\omega_1\|\varphi\|^2_{L^2(H,\nu)}, \quad \varphi \in L_0^2(H, \nu). \tag{3.73}$$

This yields (3.72) by the Hille–Yosida theorem. Finally, let us prove (3.71). Note that by (3.73) it follows that, for any $\varphi \in L_0^2(H, \nu)$,

$$\int_H |P_t\varphi|^2 d\nu \le e^{-2\omega_1 t}\int_H |\varphi|^2 d\nu.$$

Therefore for any $\varphi \in L^2(H, \nu)$ we have

$$\int_H |P_t\varphi - \overline{\varphi}|^2 d\nu = \int_H |P_t(\varphi - \overline{\varphi})|^2 d\nu \le e^{-2\omega_1 t}\int_H |\varphi - \overline{\varphi}|^2 d\nu$$

$$= e^{-2\omega_1 t}\left[\int_H |\varphi|^2 d\nu - \overline{\varphi}^2\right] \le e^{-2\omega_1 t}\int_H |\varphi|^2 d\nu.$$

The proof is complete. $\qquad\square$

We prove now the log-Sobolev inequality.

**Proposition 3.30.** *For any $\varphi \in W^{1,2}(H, \nu)$ we have*

$$\int_H \varphi^2 \log(\varphi^2) d\nu \le \frac{1}{\omega_1}\int_H |D\varphi|^2 d\nu + |\varphi|^2_{L^2(H,\nu)}\log(|\varphi|^2_{L^2(H,\nu)}). \tag{3.74}$$

*Proof.* It is enough to take $\varphi \in \mathscr{E}_A(H)$ such that $\varphi(x) \geq \varepsilon > 0$, $x \in H$. In this case we have

$$\frac{d}{dt} \int_H P_t(\varphi^2) \, \log(P_t(\varphi^2)) d\nu = \int_H K_2 P_t(\varphi^2) \log(P_t(\varphi^2)) d\nu$$

$$+ \int_H K_2 P_t(\varphi^2) d\nu.$$

The second term vanishes, due to the invariance of $\nu$. For the first term we use the identity

$$\int_H (K_2\varphi) g'(\varphi) d\nu = -\frac{1}{2} \int_H g''(\varphi) |D\varphi|^2 d\nu,$$

with $g'(\xi) = \log \xi$. Therefore we have

$$\frac{d}{dt} \int_H P_t(\varphi^2) \log(P_t(\varphi^2)) d\nu = -\frac{1}{2} \int_H \frac{1}{P_t(\varphi^2)} |DP_t(\varphi^2)|^2 d\nu. \qquad (3.75)$$

Since

$$DP_t(\varphi^2)(x) = 2\mathbb{E}[\varphi(X(t,x)) D\varphi(X(t,x)) \cdot X_x(t,x)],$$

it follows, from the Hölder inequality and (3.47), that

$$|DP_t(\varphi^2)(x)|^2 \leq 4e^{-2t\omega_1} \mathbb{E}[\varphi^2(X(t,x))] \mathbb{E}[|D\varphi|^2(X(t,x))],$$

which yields

$$|DP_t(\varphi^2)(x)|^2 \leq 4e^{-2t\omega_1} P_t(\varphi^2)(x) \, P_t(|D\varphi|^2)(x).$$

Substituting in (3.75) yields

$$\frac{d}{dt} \int_H P_t(\varphi^2) \log(P_t(\varphi^2)) d\nu \geq -2e^{-2t\omega_1} \int_H P_t(|D\varphi|^2) d\nu$$

$$= -2e^{-2t\omega_1} \int_H |D\varphi|^2 d\nu,$$

due to the invariance of $\nu$. Integrating in $t$ gives

$$\int_H P_t(\varphi^2) \log(P_t(\varphi^2)) d\nu \geq \frac{1}{\omega_1} (1 - e^{-2t\omega_1}) \int_H |D\varphi|^2 d\nu.$$

Finally, letting $t$ tend to $+\infty$, and recalling (3.55), yields

$$|\varphi|_{L^2(H,\nu)}^2 \log(|\varphi|_{L^2(H,\nu)}^2) - \int_H \varphi^2 \log(\varphi^2) d\nu \geq -\frac{1}{\omega_1} \int_H |D\varphi|^2 d\nu$$

and the conclusion follows.                                                      $\square$

## 3.7 Comparison of $\nu$ with a Gaussian measure

In this section we assume that Hypothesis 3.14 holds and in addition that $C = I$ and $F \in C_b(H; H)$.

We set again $\omega_1 := -(\omega + \kappa)$ and denote by $\nu$ the invariant measure of the transition semigroup $P_t$ defined by (3.5) (it is unique by Theorem 3.17).

We recall that, under these assumptions, the Ornstein–Uhlenbeck semigroup is strong Feller and its infinitesimal generator $L$ in $C_b(H)$ has a unique invariant measure $\mu = N_{Q_\infty}$ where

$$Q_\infty x = \int_0^{+\infty} e^{tA} e^{tA^*} x \, dt, \quad x \in H.$$

We are going to show that $\nu$ is absolutely continuous with respect to $\mu$.

### 3.7.1 First method

We follow here [35]. Let us consider the Kolmogorov operator $K$ in $C_b(H)$ defined as

$$K\varphi(x) = L\varphi(x) + \langle F(x), D\varphi(x) \rangle, \quad \varphi \in D(L, C_b(H)). \tag{3.76}$$

This definition is meaningful since, in view of (2.34), $D(L, C_b(H)) \subset C_b^1(H)$. It is convenient to introduce the approximating operators

$$K_{n,k}\varphi(x) = L_{n,k}\varphi(x) + \langle F(x), D\varphi(x) \rangle, \quad \varphi \in D(L, C_b(H)),$$

where $L_{n,k}$ is defined as

$$L_{n,k}\varphi(x) = \frac{1}{2} \operatorname{Tr} \left[ P_n D^2 \varphi(x) \right] + \langle A_k x, D\varphi(x) \rangle, \quad \varphi \in D(L_{n,k}, C_b(H)),$$

$P_n$ is an orthogonal projector of dimension $n$ and $A_k$ are the Yosida approximations of $A$.

First we show that $K$ is $m$-dissipative in $C_b(H)$ and we find an explicit formula for the resolvent $(\lambda - K)^{-1}$ of $K$.

**Lemma 3.31.** $K$ is $m$-dissipative in $C_b(H)$. Moreover, for any $\lambda > \pi \|F\|_0^2$ we have

$$(\lambda - K)^{-1} = (\lambda - L)^{-1}(1 - T_\lambda)^{-1}, \tag{3.77}$$

*where*

$$T_\lambda \varphi = \langle F, D(\lambda - L)^{-1} \varphi \rangle, \quad \varphi \in C_b(H).$$

*Finally,*

$$(\lambda - K)^{-1} f(x) = \int_0^{+\infty} e^{-\lambda t} P_t f(x) dt, \quad x \in H, \; f \in C_b(H). \tag{3.78}$$

*Proof.* Given $f \in C_b(H)$ and $\lambda > 0$, consider the equation

$$\lambda\varphi - L\varphi - \langle F(x), D\varphi \rangle = f.$$

Setting $\lambda\varphi - L\varphi = \psi$, this equation becomes $\psi - T_\lambda \psi = f$. But by (2.34) we have that

$$\|T_\lambda \psi\|_0 \leq \sqrt{\frac{\pi}{\lambda}}\, \|F\|_0.$$

So, if $\lambda > \pi\|F\|_0^2$ we have that $\|T_\lambda\psi\|_0 < 1$ and from the contraction principle it follows that $(\pi\|F\|_0^2, +\infty)$ belongs to the resolvent of $K$ and (3.77) holds. Let us show (3.78). Consider the equation

$$\lambda\varphi_{n,k} - L_{n,k}\varphi_{n,k} - \langle F(x), D\varphi_{n,k} \rangle = f,$$

which has a unique solution $\varphi_{k,n}$ for $\lambda > \pi\|F\|_0^2$, arguing as before. Moreover, by standard fixed points arguments we have, taking into account Theorem 3.5, that

$$\lim_{k\to\infty} \lim_{n\to\infty} \varphi_{k,n} = \varphi \quad \text{in } C_b(H).$$

Finally, it is easy to see that

$$(\lambda - K_{n,k})^{-1} f(x) = \int_0^{+\infty} P_t^{n,k} \varphi(x)\,dt, \quad x \in H,\ f \in B_b(H),$$

where $P_t^{n,k}$ is defined by (3.23). Thus (3.78) follows. Since, by (3.78),

$$\|(\lambda - K)^{-1}\varphi\|_0 \leq \frac{1}{\lambda}\, \|f\|_0, \quad \lambda > \pi\|F\|_0^2,$$

it follows, by a well-known property of dissipative operators, that $K$ is $m$-dissipative.  $\square$

**Remark 3.32.** In a similar way one can show that $K$ is an $m$-dissipative operator in $B_b(H)$ and that (3.78) holds for any $f \in B_b(H)$.

**Theorem 3.33.** *Assume that Hypothesis 3.14 holds, that $C = I$ and $F \in C_b(H; H)$. Then for any $f \in B_b(H)$ we have*

$$\int_H f\,d\mu = \int_H f\,d\nu + \int_H \langle F, DL^{-1}f \rangle\,d\nu, \tag{3.79}$$

*where $DL^{-1}f$ is defined by (2.38). Moreover $\nu$ is absolutely continuous with respect to $\mu$.*

*Proof.* By Lemma 3.31 we have, for any $\lambda > 0$,

$$\lambda(\lambda - L)^{-1}f = \lambda(\lambda - K)^{-1}f - \lambda(\lambda - K)^{-1}\left[\langle F, D(\lambda - L)^{-1}f \rangle\right]. \tag{3.80}$$

Now, by (2.37) we have

$$\lim_{\lambda \to 0} \lambda(\lambda - L)^{-1} f(x) = \int_H f d\mu, \quad x \in H$$

and by a similar argument, using (3.55),

$$\lim_{\lambda \to 0} \lambda(\lambda - K)^{-1} f(x) = \int_H f d\nu, \quad x \in H.$$

Thus, (3.79) follows.

Let us now prove that $\nu << \mu$. Let $\Gamma \subset H$ be a Borel set such that $\mu(\Gamma) = 0$. Then we have

$$R_t 1_\Gamma(x) = N_{e^{tA}x, Q_t}(\Gamma) = 0, \text{ for all } t > 0 \text{ and } x \in H.$$

This follows from the fact that, since $R_t$ is strong Feller, the measure $N_{e^{tA}x, Q_t}$ is absolutely continuous with respect to $\mu$. Consequently, $D(\lambda - L)^{-1} 1_\Gamma(x) = 0$ for all $x \in H$. Thus, by (3.79) it follows that $\nu(\Gamma) = \mu(\Gamma) = 0$. $\qquad \square$

### 3.7.2  Second method

Let us extend the Kolmogorov operator $K$ defined by (3.76) to $L^2(H, \mu)$ setting

$$K\varphi(x) = L_2\varphi + \langle F(x), D\varphi \rangle, \quad \varphi \in D(L_2).$$

We still denote by $K$ this extension. The reader should not confuse $K$ with $K_2$, since $K$ is defined in $L^2(H, \mu)$ whereas $K_2$ is defined in $L^2(H, \nu)$. The proof of the following lemma is very similar to that of Lemma 3.31 so it will be omitted.

**Lemma 3.34.** *The resolvent set $\rho(K)$ of $K$ includes $(\pi \|F\|_0^2, +\infty)$ and we have*

$$(\lambda - K)^{-1} = (\lambda - L_2)^{-1}(1 - T_\lambda)^{-1},$$

*where*

$$T_\lambda \varphi = \langle F, D(\lambda - L_2)^{-1}\varphi \rangle, \quad \varphi \in B_b(H).$$

*Moreover, $(\lambda - K)^{-1}$ is compact in $L^2(H, \mu)$. Finally*

$$(\lambda - K)^{-1} f(x) = \int_0^{+\infty} P_t f(x) dt, \quad x \in H, \ f \in B_b(H). \qquad (3.81)$$

By Lemma 3.34 it follows that $K$ is the infinitesimal generator of a strongly continuous semigroup which we denote by $e^{tK}$. Note that $e^{tK}\varphi = P_t\varphi$ for all $\varphi \in C_b(H)$.

We now consider the adjoint semigroup $e^{tK^*}$. We denote by $\Sigma^*$ the set of all its stationary points:

$$\Sigma^* = \left\{ \varphi \in L^2(H, \mu) : \ e^{tK^*}\varphi = \varphi, \ t \geq 0 \right\}.$$

**Lemma 3.35.** $e^{tK^*}$ *has the following properties:*

(i) *For all* $\varphi \geq 0$ *we have* $e^{tK^*}\varphi \geq 0$ $\mu$-*a.e.*

(ii) $\Sigma^*$ *is a lattice, that is if* $\varphi \in \Sigma^*$, *then* $|\varphi| \in \Sigma^*$.

*Proof.* Let $\psi_0 \geq 0$ $\mu$-a.e. Then for all $\varphi \geq 0$ and all $t > 0$ we have

$$\int_H (e^{tK}\varphi)\psi_0 \, d\mu = \int_H \varphi(e^{tK^*}\psi_0) \, d\mu \geq 0.$$

This implies that $e^{tK^*}\psi_0 \geq 0$ $\mu$-a.e, and (i) is proved.

Let us prove (ii). Assume that $\varphi \in \Sigma^*$, so that $\varphi(x) = e^{tK^*}\varphi(x)$. Then we have

$$|\varphi(x)| = |e^{tK^*}\varphi(x)| \leq e^{tK^*}(|\varphi|)(x), \quad x \in H.$$

We claim that

$$|\varphi(x)| = e^{tK^*}(|\varphi|)(x), \quad \mu \text{ a.s.}$$

Assume by contradiction that there is a Borel subset $I \subset H$ such that $\mu(I) > 0$ and $|\varphi(x)| < e^{tK^*}(|\varphi|)(x)$, $x \in I$. Then we have

$$\int_H |\varphi(x)|\mu(dx) < \int_H e^{tK^*}(|\varphi|)(x)\mu(dx).$$

On the other hand

$$\int_H e^{tK^*}(|\varphi|)d\mu = \langle e^{tK^*}(|\varphi|), 1 \rangle_{L^2(H,\mu)} = \langle |\varphi|, 1 \rangle_{L^2(H,\mu)} = \int_H |\varphi|d\mu,$$

which is a contradiction. $\qquad\square$

We prove now a regularity result for the domain $D(K^*)$ of $K^*$.

**Proposition 3.36.** *Assume that the semigroup* $e^{tA}$ *is analytic. Then the operator* $K$ *is variational and* $D(K^*) \subset W^{1,2}(H, \mu)$.

*Proof.* Let us first notice that, since $e^{tA}$ is analytic we have $Q_\infty(H) \subset D(A)$, see [29, Proposition 2.5], so that $AQ_\infty$ is bounded. Consequently, in view of (2.48) we can write

$$\int_H L_2\varphi \, \psi d\mu = \int_H \langle AQ_\infty D\psi, D\varphi\rangle d\mu, \quad \varphi \in D(L_2), \ \psi \in W^{1,2}(H, \mu).$$

Let us consider now the bilinear form $b: W^{1,2}(H, \mu) \times W^{1,2}(H, \mu) \mapsto \mathbb{R}$ defined as

$$b(u, v) = -\frac{1}{2}\int_H \langle Du, AQ_\infty Dv\rangle d\mu + \int_H \langle F, Du\rangle v d\mu.$$

Clearly $b$ is continuous, coercive and we have

$$b(u, v) = \int_H Ku \, v d\mu, \quad u, v \in D(L_2).$$

Thus $K$ is variational and consequently, its adjoint $K^*$ is variational as well and its domain belongs to $W^{1,2}(H, \mu)$. $\qquad\square$

The first part of the following result was proved in [50] and the last one in [11] with different proofs.

**Proposition 3.37.** *There exists a unique invariant measure $\nu$ for $e^{tK}$ (and for $P_t$) which is absolutely continuous with respect to $\mu$. Moreover, setting $\rho = \frac{d\nu}{d\mu}$ we have $\rho \in W^{1,2}(H, \mu)$ and $D \log \rho \in W^{1,2}(H, \nu; H)$.*

*Proof.* Let $\lambda > 0$ be fixed and let $\varphi_0$ be the function identically equal to 1. Clearly $\varphi_0 \in D(K)$ and we have $K\varphi_0 = 0$. Consequently $1/\lambda$ is an eigenvalue of the resolvent $(\lambda - K)^{-1}$ and

$$(\lambda - K)^{-1}\varphi_0 = \frac{1}{\lambda}\,\varphi_0.$$

Moreover $1/\lambda$ is a simple eigenvalue because $\mu$ is ergodic and the embedding $W^{1,2}(H, \mu) \subset L^2(H, \mu)$ is compact, see e.g. [50]. Since $D(L_2) \subset W^{1,2}(H, \mu)$, it follows that $(\lambda - K)^{-1}$ is compact as well for any $\lambda > 0$. Therefore $(\lambda - K^*)^{-1}$ is compact and $1/\lambda$ is a simple eigenvalue for $(\lambda - K^*)^{-1}$. Consequently there exists $\rho \in W^{1,2}(H, \mu)$ such that

$$(\lambda - K^*)^{-1}\rho = \frac{1}{\lambda}\,\rho. \tag{3.82}$$

It follows that $\rho \in D(K^*)$ and $K^*\rho = 0$. Since $\Sigma^*$ is a lattice by Lemma 3.35, $\rho$ can be chosen to be nonnegative and such that $\int_H \rho d\mu = 1$.

Now set

$$\nu(dx) = \rho(x)\mu(dx), \quad x \in H.$$

We claim that $\nu$ is an invariant measure for $P_t$. In fact taking the inverse Laplace transform in (3.82) we find

$$e^{tK^*}\rho = \rho,$$

which implies for any $\varphi \in C_b(H)$,

$$\int_H e^{tK}\varphi d\nu = \int_H (e^{tK}\varphi)\rho d\mu = \int_H \varphi(e^{tK^*})\rho d\mu = \int_H \varphi d\nu,$$

so that $\nu$ is invariant.

Let us prove the last statement. Since we know that $\rho \in W^{1,2}(H, \mu)$, it remains to show that $D \log \rho \in W^{1,2}(H, \nu)$. If $\varphi \in \mathscr{E}_A(H)$, then we have

$$\begin{aligned}
\int_H L_2\varphi d\nu &= \int_H L_2\varphi\, \rho d\mu = -\frac{1}{2}\int_H \langle D\varphi, D\rho\rangle d\mu \\[2mm]
&= -\frac{1}{2}\int_H \langle D\varphi, D\log\rho\rangle d\nu.
\end{aligned}$$

Consequently, since $\int_H K\varphi d\nu = 0$, we have that

$$\int_H \langle F, D\varphi\rangle d\nu = \frac{1}{2}\int_H \langle D\varphi, D\log\rho\rangle d\nu.$$

Setting $\varphi = \log \rho$ yields

$$\int_H |D \log \rho|^2 d\nu = 2 \int_H \langle F, D \log \rho \rangle d\nu$$

and the conclusion follows from the Hölder inequality.    $\square$

By Proposition 3.37 we know that the density $\rho$ belongs to $W^{1,2}(H, \mu)$. In order to find additional regularity results for $\rho$ we need more information on the domain of $K^*$. We are going to show that when $F$ is sufficiently regular, an explicit expression for $K^*$ can be found, see [50].

Let us denote by $\{e_k\}$ a complete orthonormal system in $H$ such that

$$Q_\infty e_k = \lambda_k e_k, \quad k \in \mathbb{N},$$

where $\{\lambda_k\}$ are the eigenvalues of $Q_\infty$. Set $x_k = \langle x, e_k \rangle$, $k \in \mathbb{N}$, and denote by $D_k$ the derivative in the direction of $e_k$. By (1.23) it follows that,

$$\int_H \alpha D_k \beta d\mu = -\int_H \beta D_k \alpha d\mu + \frac{1}{\lambda_k} \int_H x_k \alpha \beta d\mu, \quad k \in \mathbb{N}, \qquad (3.83)$$

where $\alpha, \beta \in C_b^1(H)$.

Let $F \in C_b^1(H; H)$. We say that $F$ has finite *divergence* if for any $x \in H$ the series

$$\operatorname{div} F(x) := \sum_{k=1}^\infty D_k F_k(x),$$

where $F_k(x) = \langle F(x), e_k \rangle$, is convergent and moreover $\operatorname{div} F \in C_b(H)$. If in addition the function

$$Q_\infty(H) \to \mathbb{R}, \quad x \to \langle Q_\infty^{-1} x, F(x) \rangle,$$

is uniquely extendible to a uniformly continuous and bounded function, we say that $F$ has finite divergence with respect to $\mu$, and we set

$$\operatorname{div}_\mu F(x) = \operatorname{div} F(x) - \langle Q_\infty^{-1} x, F(x) \rangle, \quad x \in H.$$

The following result is an easy consequence of (3.83).

**Lemma 3.38.** *Assume that $F \in C_b^1(H; H)$ has finite divergence with respect to $\mu$. Then for any $\varphi, \psi \in W^{1,2}(H, \mu)$ we have*

$$\int_H \langle F, D\varphi \rangle \psi d\mu = -\int_H \varphi \langle F, D\psi \rangle d\mu - \int_H \operatorname{div}_\mu F \varphi \, \psi \, d\mu. \qquad (3.84)$$

*Proof.* It is enough to consider $\varphi, \psi \in \mathscr{E}_A(H)$. In this case we have from (3.83),

$$\int_H \langle F, D\varphi \rangle \psi d\mu = \sum_{k=1}^{\infty} \int_H F_k D_k \varphi \, \psi d\mu$$

$$= -\sum_{k=1}^{\infty} \int_H \varphi \left( D_k F_k \psi + F_k D_k \psi \right) d\mu + \sum_{k=1}^{\infty} \frac{1}{\lambda_k} \int_H x_k F_k \varphi \psi d\mu$$

$$= -\int_H \varphi \psi \, \text{div} \, F \, d\mu - \int_H \langle F, D\psi \rangle \varphi d\mu + \int_H \langle Q_\infty^{-1} x, F \rangle \varphi \psi d\mu,$$

and the conclusion follows. $\qquad \qquad \square$

**Proposition 3.39.** *Assume that $F \in C_b^1(H; H)$ has finite divergence with respect to $\mu$. Then $D(K^*) = D(L_2)$ and*

$$K^* \psi = L_2 \psi - \langle F(x), D\psi \rangle - \text{div}_\mu \, F(x) \psi, \quad \psi \in D(K^*) = D(L_2).$$

*Proof.* Let $\varphi, \psi \in D(L_2)$. Then, taking into account (2.50) and (3.83), we have

$$\int_H K\varphi \, \psi d\mu = \int_H L_2 \varphi \, \psi d\mu + \int_H \langle F, D\varphi \rangle \psi d\mu$$

$$= \int_H \varphi \, L_2 \psi d\mu - \int_H \varphi \langle F, D\psi \rangle d\mu - \int_H \text{div}_\mu F \varphi \psi d\mu$$

which yields the conclusion. $\qquad \qquad \square$

Recalling Remark 2.51 we obtain the result,

**Corollary 3.40.** *Under the assumptions of Proposition 3.39 we have $\rho \in W^{2,2}(H, \mu)$ and $|(-A)^{1/2} D\rho| \in L^2(H, \mu)$.*

### 3.7.3 The adjoint of $K_2$

In this subsection we will find an explicit expression of the adjoint of $K_2$ in $L^2(H, \nu)$. By Theorem 3.33 we know that $\nu << \mu$ and by Proposition 3.37 that $D \log \rho \in L^2(H, \nu)$ where $\rho = \frac{d\nu}{d\mu}$.

**Proposition 3.41.** *Assume that $F \in C_b^1(H; H)$ has finite divergence with respect to $\mu$. Then we have*

$$K_2^* \varphi = L\varphi + \langle D \log \rho - F, D\varphi \rangle, \quad \varphi \in C_b^1(H). \tag{3.85}$$

*Proof.* We first notice that the measure $\nu$ is invariant for the adjoint semigroup $P_t^*$ in $L^2(H, \nu)$ since $P_t 1 = 1$, and so

$$\int_H P_t^* \varphi d\nu = \int_H \varphi d\nu, \quad \varphi \in C_b(H).$$

Now, if $\varphi, \psi \in C_b^1(H)$, we have recalling that $L_2$ is symmetric,

$$\int_H L_2 \varphi \, \psi d\nu = \int_H L_2 \varphi \, (\psi \rho) d\mu = \int_H \varphi \, L_2(\psi \rho) d\mu$$

$$= \int_H \varphi \, L_2 \psi d\nu + \int_H \varphi \, \psi \, \frac{L_2 \rho}{\rho} d\nu + \int_H \varphi \langle D\psi, D \log \rho \rangle d\nu. \tag{3.86}$$

Moreover, in view of (3.84), we have

$$\int_H \langle F, D\varphi \rangle \psi d\nu = - \int_H \varphi \operatorname{div} F \, \psi d\nu - \int_H \varphi \langle F, D\psi \rangle d\nu. \tag{3.87}$$

From (3.86) and (3.87) it follows that

$$K_2^* \psi = L_2 \psi + \langle D \log \rho - F, D\psi \rangle + V\psi,$$

where

$$V = \frac{L_2 \rho}{\rho} - \operatorname{div}_\mu F - \langle D \log \rho, F \rangle.$$

Since $K_2^* 1 = 0$ we have $V = 0$ and the conclusion follows. □

The following result is an immediate corollary of Proposition 3.41 and the integration by parts formula (3.65).

**Proposition 3.42.** *$K_2$ is symmetric if and only if*

$$F = \frac{1}{2} D \log \rho. \tag{3.88}$$

*In this case we have*

$$\int_H P_t \varphi \, \psi d\nu = - \int_H \langle D\varphi, D\psi \rangle \, d\nu. \tag{3.89}$$

**Remark 3.43.** Proceeding as in the proof of Theorem 2.56, we can prove that if $P_t$ is symmetric, then it is hypercontractive.

# Chapter 4

# Reaction-Diffusion Equations

We shall consider here a stochastic heat equation perturbed by a polynomial term of odd degree $d > 1$ having negative leading coefficient (this will ensure non-explosion). We can represent this polynomial as

$$\lambda\xi - p(\xi), \quad \xi \in \mathbb{R},$$

where $\lambda \in \mathbb{R}$ and $p$ is an increasing polynomial, that is $p'(\xi) \geq 0$ for all $\xi \in \mathbb{R}$.

## 4.1  Introduction and setting of the problem

We set $H = L^2(\mathscr{O})$ where $\mathscr{O} = [0,1]^n$, $n \in \mathbb{N}$, and denote by $\partial\mathscr{O}$ the boundary of $\mathscr{O}$ [1]. We are concerned with the stochastic differential equation

$$\begin{cases} dX(t,\xi) = [\Delta_\xi X(t,\xi) + \lambda X(t,\xi) - p(X(t,\xi))]dt + BdW(t,\xi), & \xi \in \mathscr{O}, \\[2mm] X(t,\xi) = 0, & t \geq 0,\ \xi \in \partial\mathscr{O}, \\[2mm] X(0,\xi) = x(\xi), & \xi \in \mathscr{O},\ x \in H, \end{cases}$$

$$(4.1)$$

where $\Delta_\xi$ is the Laplace operator, $B \in L(H)$ and $W$ is a cylindrical Wiener process in a probability space $(\Omega, \mathscr{F}, \mathbb{P})$ in $H$. We choose, as in the previous chapters, $W$ of the form

$$W(t,\xi) = \sum_{k=1}^\infty e_k(\xi)\beta_k(t), \quad \xi \in \mathscr{O},\ t \geq 0,$$

where $\{e_k\}$ is a complete orthonormal system in $H$ and $\{\beta_k\}$ a sequence of mutually independent standard Brownian motions on a probability space $(\Omega, \mathscr{F}, \mathbb{P})$.

---

[1] We shall consider only this choice of $\mathscr{O}$ for the sake of simplicity, for more general situations see [19].

Let us write problem (4.1) as a stochastic differential equation in the Hilbert space $H$. For this we denote by $A$ the realization of the Laplace operator with Dirichlet boundary conditions

$$\begin{cases} Ax = \Delta_\xi x, & x \in D(A), \\ D(A) = H^2(\mathscr{O}) \cap H_0^1(\mathscr{O}). \end{cases} \tag{4.2}$$

$A$ is self-adjoint and possesses a complete orthonormal system of eigenfunctions, namely

$$e_k(\xi) = (2/\pi)^{n/2} \sin(\pi k_1 \xi_1) \cdots (\sin \pi k_n \xi_n), \quad \xi = (\xi_1, \dots, \xi_n) \in \mathbb{R}^n,$$

where $k = (k_1, \dots, k_n)$, $k_i \in \mathbb{N}$.

For any $x \in H$ we set $x_k = \langle x, e_k \rangle$, $k \in \mathbb{N}^n$. Notice that

$$A e_k = -\pi^2 |k|^2, \quad k \in \mathbb{N}^n, \ |k|^2 = k_1^2 + \cdots + k_n^2.$$

Therefore, we have

$$\|e^{tA}\| \le e^{-\pi^2 t}, \quad t \ge 0. \tag{4.3}$$

**Remark 4.1.** We can also consider the realization of the Laplace operators with Neumann boundary conditions

$$\begin{cases} Nx = \Delta_\xi x, & x \in D(N), \\ D(N) = \left\{ x \in H^2(\mathscr{O}) : \dfrac{\partial x}{\partial \eta} = 0 \text{ on } \partial \mathscr{O} \right\}, \end{cases}$$

where $\eta$ represents the outward normal to $\partial \mathscr{O}$. Then

$$N f_k = -\pi^2 |k|^2 f_k, \quad k \in (\mathbb{N} \cap \{0\})^n,$$

where

$$f_k(\xi) = (2/\pi)^{n/2} \cos(\pi k_1 \xi_1) \cdots (\cos \pi k_n \xi_n),$$

$k = (k_1, \dots, k_n)$, $k_i \in \mathbb{N} \cup \{0\}$ and $|k|^2 = k_1^2 + \cdots + k_n^2$.

Concerning the operator $B$, we shall assume for simplicity that $B = (-A)^{-\gamma/2}$ where $\gamma \ge 0$ will be chosen later [2].

Now, setting $X(t) = X(t, \cdot)$ and $W(t) = W(t, \cdot)$, we shall write problem (4.1) as

$$\begin{cases} dX(t) = [AX(t) + F(X(t))]dt + (-A)^{-\gamma/2} dW(t), \\ X(0) = x, \end{cases} \tag{4.4}$$

---

[2] All following results remain true taking $B = G(-A)^{-\gamma/2}$ with $G \in L(H)$.

where $F$ is the mapping

$$F : D(F) = L^{2d}(\mathcal{O}) \subset H \to H, \quad x(\xi) \mapsto \lambda \xi - p(x(\xi)).$$

It is convenient to consider the Ornstein–Uhlenbeck process

$$\begin{cases} dZ(t) = AZ(t)dt + (-A)^{-\gamma/2}dW(t), \\ Z(0) = x, \end{cases}$$

and the corresponding transition semigroup

$$R_t\varphi(x) = \mathbb{E}[\varphi(Z(t,x))], \quad \varphi \in C_b(H).$$

Notice that, thanks to (4.3), assumptions (i)–(ii) of Hypothesis 2.1 are fulfilled with $M = 1$ and $\omega = -\pi^2$. Moreover, since $C = BB^* = (-A)^{-\gamma}$ and

$$\begin{aligned} Q_t x &= \int_0^t e^{sA} BB^* e^{sA^*} x\, ds = \int_0^t (-A)^{-\gamma} e^{2tA} x\, dt \\ &= (-A)^{-(1+\gamma)}(1 - e^{2tA})x, \quad t \geq 0, \ x \in H, \end{aligned}$$

Hypothesis 2.1–(iii) holds if and only if

$$\mathrm{Tr}\,[(-A)^{-(1+\gamma)}] = \sum_{k \in \mathbb{N}^n} |k|^{2(1+\gamma)} < +\infty. \tag{4.5}$$

Obviously, (4.5) is equivalent to

$$\gamma > \frac{n}{2} - 1. \tag{4.6}$$

In this chapter we shall always assume (4.6). Notice that can take $\gamma = 0$ and $B = I$ (white noise) only for $n = 1$.

**Exercise 4.2.** Prove that $R_t$ is strong Feller if and only if

$$\gamma < 1. \tag{4.7}$$

Consequently, since $\gamma > \frac{n}{2} - 1$, $R_t$ is strong Feller for $n < 4$.

    **Hint.** Compute $Q_t^{-1/2} e^{tA} e_k$ for all $k \in \mathbb{N}^n$ and show that Hypothesis 2.24 is fulfilled.

**Proposition 4.3.** *Assume that $\gamma > \frac{n}{2} - 1$ and $m > 1$. Then $W_A(\cdot,\cdot)$ is continuous on $[0,T] \times \mathcal{O}$, $\mathbb{P}$-almost surely and*

$$\mathbb{E}\left(\sup_{(t,\xi)\in[0,T]\times\mathcal{O}} |W_A(t,\xi)|^{2m}\right) < +\infty.$$

*Proof.* Let us check that Hypothesis 2.10 is fulfilled. Condition (i) is well known, (ii) holds with $r = 2$, (iii) holds with

$$\beta_k = \pi^2|k|^2, \quad \lambda_k = (\pi^2|k|^2)^{-\gamma}, \quad k \in \mathbb{N}^n.$$

Condition (iv) is obviously fulfilled, whereas condition (v) holds for $\alpha$ such that $2\alpha < \gamma - \frac{n}{2} + 1$.

So, the conclusion follows from Theorem 2.13. $\qquad\square$

§4.2 is devoted to solve (4.4), §4.3 to Feller and strong Feller properties of the transition semigroup and §4.4 to irreducibility. In §4.5 we study existence (and uniqueness) of an invariant measure $\nu$ and in §4.6 the transition semigroup in $L^2(H,\nu)$. §4.7 is devoted to the integration by parts formula and §4.8 to show that in some interesting cases, $\nu$ is absolutely continuous with respect to the Gaussian measure related to the case $F = 0$. Finally, in §4.9 we prove (under suitable assumptions) the compactness of the embedding of $L^2(H,\nu)$ in $W^{1,2}(H,\nu)$ and in §4.10 we study the case when (4.1) is a gradient system.

## 4.2   Solution of the stochastic differential equation

We consider here problem (4.1) (equivalently (4.4)) under the following assumptions.

**Hypothesis 4.4.**

(i) *For $\lambda \in \mathbb{R}$, $p$ is a polynomial of degree $d > 1$ such that $p'(\xi) \geq 0$ for all $\xi \in \mathbb{R}$.*

(ii) *$B = (-A)^{-\gamma/2}$ with $\gamma > \frac{n}{2} - 1$.*

As noticed before, Hypothesis 4.4 implies Hypothesis 2.1.

It is convenient, following [49], to introduce two different notions of solution of (4.1).

**Definition 4.5.** (i) *Let $x \in L^{2d}(\mathscr{O})$. We say that $X \in C_W([0,T];H)$ [3] is a mild solution of problem (4.1) if $X(t) \in L^{2d}(\mathscr{O})$ for all $t \geq 0$ and it fulfills the integral equation*

$$X(t) = e^{tA}x + \int_0^t e^{(t-s)A}F(X(s))ds + W_A(t), \quad t \geq 0, \tag{4.8}$$

*where $W_A(t)$ is the stochastic convolution*

$$W_A(t) = \int_0^t e^{(t-s)A}(-A)^{-\gamma/2}dW(s), \quad t \geq 0.$$

---

[3] Recall that $C_W([0,T];H) = C_W([0,T];L^2(\Omega,\mathscr{F},\mathbb{P};H))$.

(ii) *Let $x \in H$. We say that $X \in C_W([0,T];H)$ is a generalized solution of problem (4.1) if there exists a sequence $\{x_n\} \subset L^{2d}(\mathcal{O})$, such that*

$$\lim_{n \to \infty} x_n = x \quad \text{in } H,$$

*and*

$$\lim_{n \to \infty} X(\cdot, x_n) = X(\cdot, x) \quad \text{in } C_W([0,T];H).$$

Notice that Definition 4.5–(ii) does not depend on a sequence $\{x_n\}$.

We shall denote by $X(t,x)$ both the mild and the generalized solution of (4.1).

We shall first establish existence and uniqueness of a mild solution $X(\cdot, x)$ for any $x \in L^{2d}(\mathcal{O})$, then of a generalized solution $X(\cdot, x)$ for any $x \in L^2(\mathcal{O}) = H$. Generalized solutions are important in order to define the transition semigroup on the whole $B_b(H)$ or $C_b(H)$.

A basic tool for the proof of existence is provided by the approximating problem

$$\begin{cases} dX_\alpha(t) = (AX_\alpha(t) + F_\alpha(X_\alpha(t))dt + (-A)^{-\gamma/2}dW(t), \\ \\ X_\alpha(0) = x \in H, \end{cases} \tag{4.9}$$

where for any $\alpha > 0$, $F_\alpha$ is defined by

$$F_\alpha(x)(\xi) = \lambda x(\xi) - p_\alpha(x(\xi)),$$

and $p_\alpha$ are the Yosida approximations of $p$. We recall that $p_\alpha$ are defined by

$$p_\alpha(\eta) = \frac{1}{\alpha}(\eta - J_\alpha(\eta)), \quad \eta \in \mathbb{R}, \ \alpha > 0,$$

where

$$J_\alpha(\eta) = (1 + \alpha F(\cdot))^{-1}(\eta), \quad \eta \in \mathbb{R}.$$

We shall also write

$$p(x)(\xi) = p(x(\xi)), \quad p_\alpha(x)(\xi) = p_\alpha(x(\xi)), \quad J_\alpha(x)(\xi) = J_\alpha(x(\xi)), \quad \xi \in \mathcal{O}.$$

Let us recall some properties of the Yosida approximations, for more details see [16].

**Lemma 4.6.** *Let $\alpha > 0$. Then the following properties hold.*

(i) *$J_\alpha$ is Lipschitz continuous and*

$$|J_\alpha(\xi) - J_\alpha(\eta)| \leq |\xi - \eta|, \quad \xi, \eta \in \mathbb{R}. \tag{4.10}$$

*Moreover, $p_\alpha$ is Lipschitz continuous and*

$$|p_\alpha(\xi) - p_\alpha(\eta)| \leq \frac{2}{\alpha}|\xi - \eta|, \quad \xi, \eta \in \mathbb{R}.$$

(ii) $p_\alpha(\eta) = p(J_\alpha(\eta))$,   $\eta \in \mathbb{R}$.

(iii) $|p_\alpha(\eta)| \le |p(\eta)|$,   $\eta \in \mathbb{R}$.

*Proof.* (i) Since $p$ is increasing, for any $\eta_1, \eta_2 \in \mathbb{R}$ equations

$$\xi_i + \alpha p(\xi_i) = \eta_i, \quad i = 1, 2,$$

have unique solutions $\xi_1, \xi_2$. Multiplying both sides of the identity

$$\xi_1 - \xi_2 + \alpha(p(\xi_1) - p(\xi_2)) = \eta_1 - \eta_2$$

by $\xi_1 - \xi_2$, and taking into account the monotonicity of $p$, yields

$$(\xi_1 - \xi_2)^2 \le (\xi_1 - \xi_2)(\eta_1 - \eta_2),$$

which implies (4.10).

(ii) We have

$$p_\alpha(\eta) = \frac{1}{\alpha}\left[(1 + \alpha p)J_\alpha(\eta) - J_\alpha(\eta)\right] = p(J_\alpha(\eta)), \quad \eta \in \mathbb{R}.$$

(iii) We have

$$p_\alpha(\eta) = \frac{1}{\alpha}\left(J_\alpha(\eta + \alpha p(\eta)) - J_\alpha(\eta)\right), \quad \eta \in \mathbb{R}.$$

Consequently, by (4.10), it follows that

$$|p_\alpha(\eta)| \le \frac{1}{\alpha}|\eta + \alpha p(\eta) - \eta| = |p(\eta)|, \quad \eta \in \mathbb{R}. \qquad \square$$

**Exercise 4.7.** Let $p \ge 1$. Show that the operator $p$,

$$p(x)(\xi) = p(x(\xi)), \quad x \in L^p(\mathcal{O}) \text{ (resp. } C(\overline{\mathcal{O}})), \ \xi \in \mathbb{R},$$

is monotone in $L^p(\mathcal{O})$ (resp. $C(\overline{\mathcal{O}})$) [4].

Notice that, since $p_\alpha$ is Lipschitz continuous, $F_\alpha$ is Lipschitz continuous as well. Thus, for any $\alpha > 0$, and any $x \in H$, problem (4.9) has a unique solution $X_\alpha(\cdot, x) \in C_W([0, T]; H)$ in view of Theorem 3.2.

We can prove now, following [49], the main result of this section.

**Theorem 4.8.** *Assume that Hypothesis 4.4 holds and let $T > 0$. Then the following statements hold.*

---

[4] Let $E$ be a Banach space and $F\colon D(F) \subset E \to E$. $F$ is said to be *monotone* if $|x - y| \le$ $|x - y + \alpha(F(x) - F(y))|$ for any $\alpha > 0$ and any $x, y \in D(F)$.

(i) *If $x \in L^{2d}(\mathcal{O})$, problem (4.1) has a unique mild solution $X(\cdot, x)$. Moreover for any $m \in \mathbb{N}$, there is $c_{m,p,T} > 0$ such that*

$$\mathbb{E}\left(|X(t,x)|_{L^{2d}(\mathcal{O})}^{2m}\right) \leq c_{m,p,T}\left(1 + |x|_{L^{2d}(\mathcal{O})}^{2m}\right).$$

(ii) *If $x \in H$, problem (4.1) has a unique generalized solution $X(\cdot, x)$. In both cases $\lim\limits_{\alpha \to 0} X_\alpha(\cdot, x) = X(\cdot, x)$ in $C_W([0,T]; H)$.*

(iii) *If $x \in C(\overline{\mathcal{O}})$, problem (4.1) has a unique mild solution $X(\cdot, x)$ with $X(t,x) \in C(\overline{\mathcal{O}})$ for all $t \in [0,T]$. Moreover for any $m \in \mathbb{N}$, there is $c_{m,0,T} > 0$ such that*

$$\mathbb{E}\left(\sup_{t \in [0,T]} |X(t,x)|_{C(\overline{\mathcal{O}})}^{2m}\right) \leq c_{m,0,T}\left(1 + |x|_{C(\overline{\mathcal{O}})}^{2m}\right).$$

*Proof.* (i) Let $\alpha > \beta > 0$, and assume that $X_\alpha$ and $X_\beta$ are strong solutions of (4.9) (we can always reduce to this case by a suitable approximation). Then we have

$$\frac{d}{dt}\left(X_\alpha(t,x) - X_\beta(t,x)\right) = A(X_\alpha(t,x) - X_\beta(t,x)) + \lambda(X_\alpha(t,x) - X_\beta(t,x))$$

$$-p_\alpha(X_\alpha(t,x)) + p_\beta(X_\beta(t,x)).$$

Multiplying both sides of this identity by $X_\alpha(t,x) - X_\beta(t,x)$, and recalling Lemma 4.6–(ii) and (4.3), it follows that

$$\frac{1}{2}\frac{d}{dt}|X_\alpha(t,x) - X_\beta(t,x)|^2 \leq (\lambda - \pi^2)|X_\alpha(t,x) - X_\beta(t,x)|^2$$

$$-\langle p(J_\alpha(X_\alpha(t,x))) - p(J_\beta(X_\beta(t,x))), X_\alpha(t,x) - X_\beta(t,x)\rangle.$$

Now, by the monotonicity of $p_\alpha$ it follows that

$$\langle p(J_\alpha(X_\alpha(t,x))) - p(J_\beta(X_\alpha(t,x))), J_\alpha(X_\alpha(t,x)) - J_\beta(X_\beta(t,x))\rangle \geq 0.$$

Then, recalling that $p_\alpha(x) = \frac{1}{\alpha}(x - J_\alpha(x))$, we have

$$\frac{1}{2}\frac{d}{dt}|X_\alpha(t,x) - X_\beta(t,x)|^2 \leq (\lambda - \pi^2)|X_\alpha(t,x) - X_\beta(t,x)|^2$$

$$-\langle p_\alpha(X_\alpha(t,x)) - p_\beta(X_\beta(t,x)), \alpha p_\alpha(X_\alpha(t,x)) - \beta p_\beta(X_\beta(t,x))\rangle.$$

Now, taking into account that (by 4.6–(iii)), $|p_\alpha(\xi)| \leq |p(\xi)|$, $\xi \in \mathbb{R}$, it follows that

$$|\langle p_\alpha(X_\alpha(t,x)) - p_\beta(X_\beta(t,x)), \alpha p_\alpha(X_\alpha(t,x)) - \beta p_\beta(X_\beta(t,x))\rangle|$$

$$\leq (|p(X_\alpha(t,x))| + |p(X_\beta(t,x))|)\,(\alpha|p(X_\alpha(t,x))| + \beta|p(X_\beta(t,x))|)$$

$$\leq 2\alpha \sup_{t \in [0,T]}(|p(X_\alpha(t,x))|^2 + |p(X_\beta(t,x))|^2).$$

But we have for a suitable constant $c$,

$$|p(X_\alpha(t,x))|^2_{L^2(\mathcal{O})} \le c(1 + |X_\alpha^d(t,x)|^2_{L^2(\mathcal{O})})$$

$$= c\left(1 + \int_{\mathcal{O}} (X_\alpha(t,x))^{2d} d\xi\right) = c\left(1 + |X_\alpha^d(t,x)|^{2d}_{L^{2d}(\mathcal{O})}\right).$$

Therefore

$$\frac{1}{2}\frac{d}{dt}|X_\alpha(t,x) - X_\beta(t,x)|^2 \le (\lambda - \pi^2)|X_\alpha(t,x) - X_\beta(t,x)|^2 \tag{4.11}$$

$$+ c\,\alpha\left(1 + \sup_{t\in[0,T]} (|X_\alpha^d(t,x)|^{2d}_{L^{2d}(\mathcal{O})} + |X_\beta^d(t,x)|^{2d}_{L^{2d}(\mathcal{O})})\right).$$

So, it remains to estimate $|X_\alpha^d(t,x)|^{2d}_{L^{2d}(\mathcal{O})}$. To this purpose we reduce equation (4.9) to a family of deterministic integral equations. More precisely, setting $Y_\alpha(t) = X_\alpha(t) - W_A(t)$, (4.9) becomes

$$\begin{cases} Y_\alpha'(t) = (A + \lambda)Y_\alpha(t) - p_\alpha(Y_\alpha(t) + W_A(t)) + \lambda W_A(t), \quad t \in [0,T], \\[2mm] Y_\alpha(0) = x. \end{cases}$$

Now, multiplying both sides of the first equation by $(Y_\alpha(t))^{2d-1}$ yields (after an integration by parts)

$$\frac{1}{2d}\frac{d}{dt}\int_{\mathcal{O}}|Y_\alpha(t)|^{2d} d\xi + (2d-1)\int_{\mathcal{O}}|Y_\alpha(t)|^{2d-2}|\nabla_\xi Y_\alpha(t)|^2 d\xi$$

$$= \lambda\int_{\mathcal{O}}|Y_\alpha(t)|^{2d} d\xi - \int_{\mathcal{O}}[p_\alpha(Y_\alpha(t) + W_A(t)) - p_\alpha(W_A(t))]Y_\alpha(t)^{2d-1} d\xi$$

$$+ \int_{\mathcal{O}}[-p_\alpha(W_A(t)) + \lambda W_A(t)]Y_\alpha(t)^{2d-1} d\xi.$$

Taking into account the monotonicity of $p_\alpha$, we obtain

$$\frac{1}{2d}\frac{d}{dt}\int_{\mathcal{O}}|Y_\alpha(t)|^{2d} d\xi \le \lambda\int_{\mathcal{O}}|Y_\alpha(t)|^{2d} d\xi$$

$$+ \int_{\mathcal{O}}[-p_\alpha(W_A(t)) + \lambda W_A(t)]Y_\alpha(t)^{2d-1} d\xi.$$

By the Hölder inequality, it follows that

$$\frac{1}{2d}\frac{d}{dt}\int_{\mathcal{O}}|Y_\alpha(t)|^{2d} d\xi \le \lambda\int_{\mathcal{O}}|Y_\alpha(t)|^{2d} d\xi$$

$$+ \left(\int_{\mathcal{O}}[-p_\alpha(W_A(t)) + \lambda W_A(t)]^{2d} d\xi\right)^{\frac{1}{2d}} \left(\int_{\mathcal{O}}|Y_\alpha(t)|^{2d} d\xi\right)^{\frac{2d-1}{2d}}.$$

Using the standard inequality

$$ab \leq \frac{1}{2d} a^{2d} + \frac{2d-1}{2d} b^{\frac{2d}{2d-1}}, \quad a, b > 0,$$

yields

$$\frac{1}{2d} \frac{d}{dt} \int_{\mathcal{O}} |Y_\alpha(t)|^{2d} d\xi \leq \left( \lambda + \frac{2d-1}{2d} \right) \int_{\mathcal{O}} |Y_\alpha(t)|^{2d} d\xi$$

$$+ \frac{1}{2d} \int_{\mathcal{O}} [p_\alpha(W_A(t)) + \lambda W_A(t)]^{2d} d\xi$$

$$\leq \left( \lambda + \frac{2d-1}{2d} \right) \int_{\mathcal{O}} |Y_\alpha(t)|^{2d} d\xi + c_1 \int_{\mathcal{O}} (1 + |W_A(t)|^{2d^2}) d\xi,$$

for a suitable constant $c_1$ independent of $\alpha$. Integrating with respect to $t$, yields

$$|Y_\alpha(t)|^{2d}_{L^{2d}(\mathcal{O})} \leq e^{(2d\lambda + 2d - 1)t} |x|^{2d}_{L^{2d}(\mathcal{O})}$$

$$+ c_1 |\mathcal{O}| \int_0^t e^{(2d\lambda + 2d - 1)(t-s)} \left( 1 + \sup_{s \in [0,T]} |W_A(s,\xi)|^{2d^2} \right) ds,$$

where $|\mathcal{O}|$ is the Lebesgue measure of $\mathcal{O}$. Consequently there exists a positive constant $c(T,d)$ such that, for $t \in [0,T]$,

$$|Y_\alpha(t)|^{2d}_{L^{2d}(\mathcal{O})} \leq c(T,d) \left( 1 + |x|^{2d}_{L^{2d}(\mathcal{O})} + \sup_{s \in [0,T], \xi \in \mathcal{O}} |W_A(s,\xi)|^{2d^2} \right). \tag{4.12}$$

By Proposition 4.3 there exists $c_2(T,d) > 0$ such that

$$\mathbb{E} \left( |Y_\alpha(t)|^{2d}_{L^{2d}(\mathcal{O})} \right) \leq c_2(T,d) \left( |x|^{2d}_{L^{2d}(\mathcal{O})} + 1 \right),$$

and consequently there exists $c_3(T,d) > 0$ such that

$$\mathbb{E} \left( |X_\alpha(t,x)|^{2d}_{L^{2d}(\mathcal{O})} \right) \leq c_4(T,d) \left( |x|^{2d}_{L^{2d}(\mathcal{O})} + 1 \right), \quad t \in [0,T]. \tag{4.13}$$

Now, taking expectation in (4.11) and taking into account (4.12), we find that for a suitable constant $C_1$,

$$\frac{1}{2} \frac{d}{dt} \mathbb{E} \left( |X_\alpha(t,x) - X_\beta(t,x)|^2 \right) \leq C_1 \, \alpha \left( |x|^{2d}_{L^{2d}(\mathcal{O})} + 1 \right).$$

This shows that the sequence $\{X_\alpha\}$ is Cauchy in $C_W([0,T]; H)$. Let

$$X = \lim_{\alpha \to 0} X_\alpha \quad \text{in } C_W([0,T]; H).$$

Using (4.13), we can pass to the limit in the equation

$$X_\alpha(t) = e^{tA}x + \int_0^t e^{(t-s)A}F_\alpha(X_\alpha(s))ds + W_A(t), \quad t \geq 0,$$

and we find that $X$ fulfills (4.8). Existence is proved.

Let us prove uniqueness. Let $X_i$, $i = 1, 2$ be two mild solutions of (4.8). Then $X_1 - X_2$ is the mild solution of the deterministic problem

$$\begin{cases} \dfrac{d}{dt}(X_1 - X_2) = A(X_1 - X_2) + F(X_1) - F(X_2), \\ \\ (X_1 - X_2)(0) = 0. \end{cases}$$

Multiplying both sides of the first identity by $(X_1 - X_2)$, integrating with respect to $\xi$ and taking into account the monotonicity of $p$, yields

$$\frac{d}{dt}|X_1 - X_2|^2 \leq (\lambda - \pi^2)|X_2 - X_1|^2,$$

which implies $X_1 = X_2$. So part (i) is proved.

Let us prove (ii). Let $x \in H$ and let $\{x_k\} \subset L^{2d}(\mathscr{O})$ be such that $x_n \to x$ in $H$. Let $X_n = X(\cdot, x_n)$ be the mild solution corresponding to $x_n$. Then $X_n - X_m$ is the mild solution of the deterministic problem

$$\begin{cases} \dfrac{d}{dt}(X_n - X_m) = A(X_n - X_m) + F(X_n) - F(X_m), \\ \\ (X_n - X_m)(0) = x_n - x_m. \end{cases}$$

It follows that

$$\frac{d}{dt}|X_n - X_m|^2 \leq (\lambda - \pi^2)|X_n - X_m|^2,$$

which yields

$$|X_n - X_m|^2 \leq e^{2(\lambda - \pi^2)}|x_n - x_m|^2.$$

Therefore the sequence $\{X_n\}$ is Cauchy and consequently it converges to the generalized solution of problem (4.8). The uniqueness of the generalized solution can be proved as before.

Finally, (iii) can be proved in an analogous way using the dissipativity of $p$ in $C(\overline{\mathscr{O}})$, see [49] for details. $\qquad\square$

## 4.3 Feller and strong Feller properties

We assume here that Hypothesis 4.4 holds. For any $x \in H$ let $X(t,x)$ be the generalized solution of (4.1) and, for any $\alpha > 0$, let $X_\alpha(t,x)$ be the mild solution of (4.9). We recall that

$$\lim_{\alpha \to 0} X_\alpha(\cdot, x) = X(\cdot, x) \quad \text{in } C_W([0,T]; H), \quad x \in H.$$

Let us define the *transition semigroups*,

$$P_t\varphi(x) = \mathbb{E}[\varphi(X(t,x))], \quad t \ge 0, \ \varphi \in B_b(H),$$

and

$$P_t^\alpha \varphi(x) = \mathbb{E}[\varphi(X_\alpha(t,x))], \quad t \ge 0, \ \varphi \in B_b(H).$$

**Proposition 4.9.** *$P_t$ is a one-parameter semigroup on $B_b(H)$. Moreover, $P_t$ is Feller.*

*Proof.* Let $t, s \ge 0$ and $\varphi \in C_b(H)$. Since we have (see Proposition 3.9)

$$P_t^\alpha P_s^\alpha \varphi = P_{t+s}^\alpha \varphi, \quad \alpha > 0,$$

letting $\alpha \to 0$ we see that $P_t$ is a one-parameter semigroup on $B_b(H)$.

Let us show now that $P_t$ maps $C_b(H)$ into itself. For this it is enough to prove that $P_t\varphi \in C_b(H)$ for all $\varphi \in C_b^1(H)$, since $C_b^1(H)$ is dense in $C_b(H)$, see Theorem 1.1. Let $\varphi \in C_b^1(H)$ and $x_0, x \in H$. Then we have

$$|P_t\varphi(x) - P_t\varphi(x_0)| \le \|\varphi\|_1 \left( \mathbb{E}[|X(t,x) - X(t,x_0)|^2] \right)^{1/2}.$$

Since $X \in C_W([0,T]; H)$ the conclusion follows. $\square$

We prove now, following [19] the strong Feller property of $P_t$.

**Lemma 4.10.** *Assume that $n/2 - 1 < \gamma < 1$ [5] and let $\varphi \in C_b(H)$. Then for any $T > 0$ there exists $C_T > 0$ such that*

$$\|DP_t^\alpha \varphi\|_0 \le C_T \|\varphi\|_0, \quad \alpha > 0, \tag{4.14}$$

*Proof.* For any $h \in H$ we set $\eta_\alpha^h(t,x) = DX_\alpha(t,x) \cdot h$. Then by Theorem 3.6 it follows that $\eta_\alpha^h(t,x)$ is the mild solution of the equation

$$\begin{cases} \dfrac{d}{dt} \eta_\alpha^h(t,x) &= (A + \lambda)\eta_\alpha^h(t,x) - p_\alpha'(X_\alpha(t,x)) \cdot \eta_\alpha^h(t,x), \\[2mm] \eta^h(0,x) &= h. \end{cases}$$

---

[5] This obviously implies $n < 4$.

Multiplying the first equation by $\eta_\alpha^h(t, x)$, integrating by parts, and taking into account that $p'_\alpha \geq 0$, yields

$$\frac{1}{2} \frac{d}{dt} |\eta_\alpha^h(t, x)|^2 + \int_{\mathcal{O}} |D_\xi \eta_\alpha^h(t, x)|^2 d\xi \leq \lambda |\eta_\alpha^h(t, x)|^2. \tag{4.15}$$

Recalling that by the classical Poincaré inequality we have

$$\int_{\mathcal{O}} |D_\xi \eta_\alpha^h(t, x)|^2 d\xi \geq \pi^2 \int_{\mathcal{O}} |\eta_\alpha^h(t, x)|^2 d\xi, \tag{4.16}$$

we deduce that

$$\frac{1}{2} \frac{d}{dt} |\eta_\alpha^h(t, x)|^2 \leq (\lambda - \pi^2) |\eta_\alpha^h(t, x)|^2,$$

which yields

$$|\eta_\alpha^h(t, x)| \leq e^{(\lambda - \pi^2)t} |h|, \quad x \in H, \ t \geq 0. \tag{4.17}$$

Now, by (4.15) and (4.17) it follows that there exists $C_{1,T} > 0$ such that

$$\int_0^T |(-A)^{1/2} \eta_\alpha^h(t, x)|^2 dt \leq C_{1,T} |h|^2, \quad x \in H. \tag{4.18}$$

Let now consider the Bismut–Elworthy formula (3.34),

$$\langle DP_t^\alpha \varphi(x), h \rangle = \frac{1}{t} \, \mathbb{E} \left[ \varphi(X_\alpha(t, x)) \int_0^t \langle (-A)^{\frac{\gamma}{2}} \eta^h(s, x), dW(s) \rangle \right].$$

Using the Hölder inequality, we find that, for all $T > 0$,

$$\|DP_T^\alpha \varphi(x)\|_0^2 \leq T^{-2} \|\varphi\|_0 \mathbb{E} \int_0^T |(-A)^{\frac{\gamma}{2}} \eta_\alpha^h(t, x)|^2 dt.$$

Notice that, since $\gamma < 1$ the operator $(-A)^{\frac{\gamma}{2}} (-A)^{-\frac{1}{2}} = (-A)^{\frac{\gamma-1}{2}}$ is bounded with norm equal to $\pi^{2-2\gamma}$. Now, the conclusion follows from (4.18). $\qquad \square$

**Theorem 4.11.** *Assume that Hypothesis 4.4 holds with $\gamma < 1$. Then $P_t$ is strong Feller.*

*Proof.* Let $T > 0$, $\varphi \in C_b(H)$. By Lemma 4.10, letting $\alpha \to 0$, it follows that

$$|P_T \varphi(x) - P_T \varphi(y)| \leq C_T \|\varphi\|_0 |x - y|, \quad x, y \in H.$$

Now the conclusion follows arguing as in the proof of Step 2 of Theorem 3.11. $\qquad \square$

## 4.4  Irreducibility

In this section we prove that, under Hypothesis 4.4, $P_t$ is irreducible. We shall use the notation

$$\| \cdot \| = \| \cdot \|_{C(\overline{\mathcal{O}})}$$

and

$$M(R) = \sup_{|\xi| \leq R} p'(\xi).$$

We follow [19]. The first basic tool for proving irreducibility of $P_t$ is, as we have seen in Chapter 3, the approximate controllability of system

$$\begin{cases} y'(t) = Ay(t) + F(y(t)) + (-A)^{-\gamma/2} u(t), \\ y(0) = x. \end{cases} \tag{4.19}$$

We denote by $y(t, x; u)$ the mild or generalized solution of (4.19), which can be defined as in Definition 4.5.

We say that system (4.19) is *approximatively controllable* in time $T > 0$ if for any $x_0, x_1 \in H$ and any $\varepsilon > 0$ there exists $u \in C([0, T]; C(\overline{\mathcal{O}}))$ [6] such that

$$|y(T, x_0; u)| < \varepsilon. \tag{4.20}$$

**Proposition 4.12.** *System* (4.19) *is approximatively controllable in any time* $T > 0$.

*Proof.* We shall need the following notation. For any $z_0, z_1 \in \{x \in C^2(\overline{\mathcal{O}}) : x = 0 \text{ on } \partial\mathcal{O}\}$ we set

$$\alpha_{z_0, z_1}(t) = \frac{T - t}{T} z_0 + \frac{t}{T} z_1, \quad t \in [0, T],$$

and

$$\beta_{z_0, z_1}(t) = \frac{d}{dt} \alpha_{z_0, z_1}(t) - A\alpha_{z_0, z_1}(t) - F(\alpha_{z_0, z_1}(t)), \quad t \in [0, T].$$

Note that $\beta_{z_0, z_1}(t) \in C(\overline{\mathcal{O}})$ for all $t \in [0, T]$ and we have

$$\alpha_{z_0, z_1}(0) = z_0, \quad \alpha_{z_0, z_1}(T) = z_1.$$

Let now $x_0, x_1 \in H$ and $\varepsilon > 0$ be fixed. Choose $z_0, z_1 \in \{x \in C^2(\overline{\mathcal{O}}) : x = 0 \text{ on } \partial\mathcal{O}\}$ and $u \in C([0, T], C(\overline{\mathcal{O}}))$ such that

(i) $|x_0 - z_0| < c\varepsilon, \quad |x_1 - z_1| < c\varepsilon,$

(ii) $|\beta_{z_0, z_1}(t) - (-A)^{-\gamma/2} u(t)| \leq c\varepsilon, \quad t \in [0, T],$

---

[6] The usual definition requires only $u \in L^2(0, T; H)$, but we shall need $u \in C([0, T]; C(\overline{\mathcal{O}}))$ in what follows.

where $c$ will be made precise later. We are going to show that the constant $c$ may be choosen such as $u$ fulfills (4.20). We have in fact

$$|y(T, x_0; u) - x_1| \leq |y(T, x_0; u) - y(T, z_0; u)| + |y(T, z_0; u) - x_1|$$

$$\leq |y(T, x_0; u) - y(T, z_0; u)| + |y(T, z_0; u) - \alpha_{z_0, z_1}(T)| = I_1 + I_2. \quad (4.21)$$

Let us estimate $I_1$. Set $v(t) = y(t, x_0; u) - y(t, z_0; u)$, $t \in [0, T]$, then $v(t)$ is a mild solution of

$$v'(t) = (A + \lambda)v(t) - p(y(t, x_0; u)) + p(y(t, z_0; u)).$$

By the monotonicity of $p$ and $A$ it follows that

$$|I_1| = |v(T)| \leq e^{\lambda T}|x_0 - z_0| \leq ce^{\lambda T}\varepsilon. \quad (4.22)$$

Let us estimate $I_2$. Set $w(t) = y(t, z_0; u) - \alpha_{z_0, z_1}(t)$, $t \in [0, T]$. Then $w$ is a mild solution of

$$w'(t) = (A + \lambda)w(t) - p(y(t, z_0; u)) + p(\alpha_{z_0, z_1}(t)) + (-A)^{-\gamma/2}\, u(t) - \beta_{z_0, z_1}(t).$$

Again by the monotonicity of $p$ and $A$ it follows that

$$|I_2| = |w(T)| \leq e^{\lambda T}\int_0^T |(-A)^{-\gamma/2}\, u(t) - \beta_{z_0, z_1}(t)|dt$$

$$\leq cTe^{\lambda T}\varepsilon. \quad (4.23)$$

So, by (4.21) we have, taking into account (4.22) and (4.23), that

$$|y(T, x_0; u) - x_1| \leq ce^{\lambda T}\varepsilon(1 + T).$$

The conclusion follows.                                                          $\square$

**Lemma 4.13.** *Let* $T > 0$, $x_0 \in C(\overline{\mathscr{O}})$, $z_i \in C([0, T], C(\overline{\mathscr{O}}))$, $i = 1, 2$ *and let* $y_i$, $i = 1, 2$ *be the solutions in* $C([0, T], C(\overline{\mathscr{O}}))$ *of the integral equations*

$$y_i(t) = e^{t(A-\lambda)}x_0 + \int_0^t e^{(t-s)(A-\lambda)}p(y_i(s))ds + z_i(t), \quad t \in [0, T], i = 1, 2. \quad (4.24)$$

*Then we have*

$$|y_1(T) - y_2(T)| \leq (1 + TM(\|z_1 - z_2\| + \|y_2\|))\|z_1 - z_2\|. \quad (4.25)$$

*Proof.* Set $U_i = y_i - z_i$, $i = 1, 2$ and $\rho = U_1 - U_2$. Then $\rho$ is the mild solution of the equation

$$\frac{d\rho}{dt} = A\rho + F(y_1(t)) - F(y_2(t)), \quad \rho(0) = 0.$$

Therefore

$$\frac{d^-}{dt} |\rho(t)| \leq \langle A\rho, \delta_\rho(t) \rangle$$

$$+\langle F(\rho(t) + z_1(t) - z_2(t) + y_2(t)) - F(z_1(t) - z_2(t) + y_2(t)), \delta_\rho(t) \rangle$$

$$+\langle F(z_1(t) - z_2(t) + y_2(t)) - F(y_2(t)), \delta_\rho(t) \rangle$$

$$\leq \langle A\rho, \delta_\rho(t) \rangle + \langle F(z_1 - z_2(t) + y_2(t)) - F(y_2(t)), \delta_\rho(t) \rangle$$

$$\leq -\pi^2 |\rho(t)| + |F(z_1 - z_2(t) + y_2(t)) - F(y_2(t))|,$$

where $\frac{d^-|\rho|}{dt}$ is the left derivative of $|\rho(t)|$ and $\delta_\rho(t) = \frac{\rho(t)}{|\rho(t)|}$. Consequently

$$|\rho(t)| \leq \int_0^t e^{-(t-s)\pi^2} |F(z_1(s) - z_2(s) + y_2(s)) - F(y_2(s))| ds$$

$$\leq TM(\|z_1 - z_2\| + \|y_2\|)\|z_1 - z_2\|.$$

Since

$$|y_1(t) - y_2(t)| \leq |\rho(t)| + |z_1(t) - z_2(t)|, \quad t \in [0, T],$$

(4.25) follows.                                                                    □

We are now in a position to prove the result,

**Proposition 4.14.** *The semigroup $P_t$ is irreducible.*

*Proof.* Let $T > 0$, $x_0, x_1 \in H$ and let $\varepsilon > 0$. We first assume in addition that $x_0, x_1 \in C(\overline{\mathcal{O}})$. We have to prove that

$$\mathbb{P}(|X(t, x_0) - x_1| \geq \varepsilon) < 1. \tag{4.26}$$

By Proposition 4.12 there exists $u \in C([0, T]; C(\overline{\mathcal{O}}))$ such that $|y(T, x_0; u) - x_1| \leq \frac{\varepsilon}{2}$. Set

$$\sigma_u(t) = \int_0^t e^{(t-s)(A-\lambda)} u(s) ds.$$

We have,

$$X(t, x_0) = e^{t(A-\lambda)} x_0 + \int_0^t e^{(t-s)(A-\lambda)} p(X(s, x_0)) ds + W_A(t), \quad t \in [0, T],$$

and

$$y(t) = e^{t(A-\lambda)} x_0 + \int_0^t e^{(t-s)(A-\lambda)} p(y(s)) ds + \sigma_u(t)(t), \quad t \in [0, T],$$

where $y(t) = y(t, x_0; u)$. Since

$$|X(T, x_0) - x_1| \leq |X(T, x_0) - y(T)| + |y(T) - x_1| + \frac{\varepsilon}{2},$$

we have obviously that

$$\mathbb{P}\left(|X(T, x_0) - x_1| \geq \varepsilon\right) \leq \mathbb{P}\left(|X(T, x_0) - y(T)| \geq \frac{\varepsilon}{2}\right).$$

But by Lemma 4.13 we have that

$$|X(T, x_0) - y(T)| \leq (1 + TM(\|W_A - \sigma_u\| + \|y\|)\|W_A - \sigma_u\|.$$

Therefore

$$\mathbb{P}(|X(t, x_0) - x_1| \geq \varepsilon) \leq \mathbb{P}(|X(t, x_0) - y(T)| \geq \varepsilon/2)$$

$$\leq \mathbb{P}\left((1 + TM(\|W_A - \sigma_u\| + \|y\|)\|W_A - \sigma_u\| \geq \frac{\varepsilon}{2}\right) < 1,$$

since $W_A(\cdot)$ is full in $C([0, T]; C(\overline{\mathcal{O}}))$, see Exercise 2.16.

It remains to examine the case when $x_0, x_1 \in H$. Since $C(\overline{\mathcal{O}})$ is dense in $H$, there exists $\overline{x}_0 \in C(\overline{\mathcal{O}})$ such that

$$B(\overline{x}_0, \varepsilon/2) \subset B(x_0, \varepsilon).$$

Then we have

$$P_t 1_{B(x_0, \varepsilon)}(x_1) \geq \mathbb{P}_t 1_{B(\overline{x}_0, \varepsilon/2)}(x_1) > 0,$$

by the previous proof.                                                                  □

## 4.5   Existence of invariant measure

We shall distinguish two cases: $\lambda \leq 0$ (in this case the nonlinear operator $F$ is dissipative) and $\lambda > 0$. Then we shall consider the special case when $\gamma = 0$. In this case $P_t$ is symmetric and (4.1) is a *gradient system*, as we shall see in §4.10. Here the invariant measure is an explicit expression.

### 4.5.1   The dissipative case

Proposition 4.15 and Theorem 4.16 below can be proved as Proposition 3.16 and Theorem 3.17 respectively, so we will omit the corresponding proofs.

**Proposition 4.15.** *Assume that Hypothesis 4.4 holds with $\lambda \leq 0$. Then there exists $\zeta \in L^2(\Omega, \mathscr{F}, \mathbb{P}; H)$ such that*

$$\lim_{s \to +\infty} X(0, -s, x) = \zeta \quad in \ L^2(\Omega, \mathscr{F}, \mathbb{P}; H), \quad x \in H.$$

**Theorem 4.16.** *Assume that Hypothesis 4.4 holds with $\lambda \le 0$. Then the semigroup $P_t$ has a unique invariant measure $\nu$. Moreover*

$$\lim_{t \to +\infty} P_t \varphi(x) = \int_H \varphi(y) \nu(dy), \quad x \in H.$$

### 4.5.2 The non-dissipative case

Here we need an assumption stronger than Hypothesis 4.4.

**Hypothesis 4.17.**

(i) $\gamma > \frac{n}{2}$.

(ii) *p is an increasing polynomial of degree $d \ge 3$.*

From (i) it follows that $\mathrm{Tr}\,[(-A)^{-\gamma}] < +\infty$ and from (ii) there exists $g > 0$ such that

$$(p(\xi) - p(\eta))(\xi - \eta) \ge g(\xi - \eta)^4.$$

In this case we are not able to show that there exists the limit

$$\lim_{s \to +\infty} X(0, -s, x) = \zeta,$$

so, in order to prove the existence of an invariant measure, we shall apply the Krylov–Bogoliubov theorem (Theorem 1.11) as in [49].

**Proposition 4.18.** *Assume that Hypothesis 4.17 holds. Then there exists an invariant measure $\nu$ for $P_t$.*

*Proof.* Let us apply the Itô formula to $\varphi(x) = \frac{1}{2}\,|x|^2$ for the process $X(t, x)$ (to be rigorous, one should establish the formula for the process $X_\alpha(t, x)$ and then let $\alpha \to 0$). We have

$$\mathbb{E}|X(t, x)|^2 = |x|^2 + \mathbb{E} \int_0^t \langle AX(s, x) + F(X(s, x)), X(s, x)\rangle ds + t\mathrm{Tr}\,[(-A)^{-\gamma}],$$

which yields

$$\frac{d}{dt}\,\mathbb{E}|X(t, x)|^2 = \mathbb{E}[\langle AX(t, x) + F(X(t, x)), X(t, x)\rangle] + \mathrm{Tr}\,[(-A)^{-\gamma}]. \qquad (4.27)$$

Setting $\|x\| = |x|_{H_0^1(0,1)}$, we have

$$\langle AX(s, x) + F(X(s, x)), X(s, x)\rangle$$

$$= \langle AX(s, x) + \lambda X(s, x) - p(X(s, x)) - p(0) + p(0), X(s, x)\rangle$$

$$= -\|X(s, x)\|^2 + \lambda|X(s, x)|^2$$

$$- \langle (p(X(s, x)) - p(0)), X(s, x)\rangle - \langle p(0), X(s, x)\rangle$$

$$\le -\|X(s, x)\|^2 - g|X(s, x)|^4 + \lambda|X(s, x)|^2 + |p(0)|\,|X(s, x)|.$$

Therefore, from (4.27) we have that, for a suitable constant $c$,

$$\frac{d}{dt}\,\mathbb{E}|X(t,x)|^2 + \|X(t,x)\|^2 \le c(1+|X(t,x)|^2) - g|X(s,x)|^4.$$

By a standard comparison result it follows that there exists $c_1 > 0$ such that $\mathbb{E}|X(t,x)|^2 \le c_1$ for all $t \ge 0$, so that

$$\mathbb{E}|X(t,x)|^2 + \int_0^t \mathbb{E}\|X(s,x)\|^2 ds \le c_1 t \quad \text{for all } t \ge 0.$$

Consequently

$$\frac{1}{t}\int_0^t \mathbb{E}\|X(s,x)\|^2 ds \le c_1 \quad \text{for all } t \ge 0. \tag{4.28}$$

Let now $x_0 \in H$ be fixed and set

$$\mu_{T,x_0} = \frac{1}{T}\int_0^T \lambda_{t,x_0} dt,$$

where $\lambda_{t,x_0}$ is the law of $X(t,x_0)$, and

$$B_R = \{x \in H_0^1(\mathcal{O}) : \|x\| \le R\}$$

for any $R > 0$. Notice that $B_R$ is a compact set in $H$ since the embedding of $H_0^1(\mathcal{O})$ into $L^2(\mathcal{O})$ is compact. Consequently,

$$\mu_{T,x_0}(B_R^c) = \frac{1}{T}\int_0^T \lambda_{t,x_0}(B_R^c)dt = \frac{1}{T}\int_0^T \mathbb{P}(\|X(t,x_0)\| \ge R)dt.$$

But

$$\mathbb{P}(\|X(t,x_0)\| \ge R) = \int_{\{\|X(t,x_0)\|\ge R\}} d\mathbb{P}$$

$$\le \frac{1}{R^2}\int_\Omega \|X(t,x_0)\|^2 d\mathbb{P} = \frac{1}{R^2}\,\mathbb{E}\|X(t,x_0)\|^2.$$

Thus, by (4.28) it follows that

$$\mu_{T,x_0}(B_R^c) \le \frac{1}{R^2}\frac{1}{T}\int_0^T \|X(t,x_0)\|^2 dt \le \frac{c}{R^2},$$

so that the family $\{\mu_{T,x_0}\}_{T>0}$ is tight. Applying the Krylov–Bogoliubov theorem (Theorem 1.11), yields the conclusion. $\qquad\square$

**Remark 4.19.** Assume that $n = 1, B = I$. Then by Proposition 4.18, $P_t$ has an invariant measure $\nu$. Moreover, by Proposition 4.12 and Theorem 4.11, $P_t$ is irreducible and strong Feller. Consequently in this case $P_t$ has a unique invariant measure in view of Theorem 1.12.

## 4.6 The transition semigroup in $L^2(H,\nu)$

In this section we shall assume for simplicity that Hypothesis 4.4 holds with $\lambda = 0$. Then the transition semigroup $P_t$ has a unique invariant measure $\nu$ by Theorem 4.16.

Proceeding as before, we see that $P_t$ can be uniquely extended to a strongly continuous semigroup of contractions in $L^2(H,\nu)$, whose infinitesimal generator we denote by $K_2$. We want to show that $K_2$ is the closure of the differential operator $K_0$ defined by

$$
\begin{aligned}
K_0\varphi &= \frac{1}{2}\,\mathrm{Tr}\,[(-A)^{-\gamma}D^2\varphi] + \langle x, A^*D\varphi\rangle + \langle F(x), D\varphi\rangle \\
&= L\varphi + \langle F(x), D\varphi\rangle, \quad \varphi \in \mathscr{E}_A(H),\ x \in H,
\end{aligned}
$$

where $L$ is the Ornstein–Uhlenbeck operator introduced in Chapter 2.

It is clear that for any $\varphi \in \mathscr{E}_A(H)$ there exist two positive contants $a, b$ (depending on $\varphi$) such that

$$
|L\varphi(x)| \le a + b|x|, \quad x \in H. \tag{4.29}
$$

So, in order that $K_0\varphi$ belongs to $L^2(H,\nu)$, we need that $F(x) \in L^2(H,\nu)$. This is provided by the following result.

**Proposition 4.20.** *Assume that Hypothesis 4.4 holds with $\lambda = 0$. Then there exists $c_d > 0$ such that*

$$
\int_H |x|^{2d}_{L^{2d}(\mathscr{O})}\nu(dx) \le c_d.
$$

*Proof.* Setting $Y_\alpha(t) = X_\alpha(t) - W_A(t)$, (4.9) reduces to

$$
\begin{cases}
Y'_\alpha(t) = AY_\alpha(t) - p_\alpha(Y_\alpha(t) + W_A(t)), & t \in [0, T], \\
Y_\alpha(0) = x.
\end{cases}
$$

Now, multiplying both sides of the first equation by $(Y_\alpha(t))^{2d-1}$ yields (after integration by parts)

$$
\frac{1}{2d}\frac{d}{dt}\int_{\mathscr{O}} |Y_\alpha(t)|^{2d}d\xi + (2d-1)\int_{\mathscr{O}} |Y_\alpha(t)|^{2d-2}|\nabla_\xi Y_\alpha(t)|^2 d\xi
$$

$$
= \int_{\mathscr{O}} [p_\alpha(Y_\alpha(t) + W_A(t)) - p_\alpha(W_A(t))]Y_\alpha(t)^{2d-1}d\xi
$$

$$
+ \int_{\mathscr{O}} p_\alpha(W_A(t))Y_\alpha(t)^{2d-1}d\xi.
$$

Taking into account the monotonicity of $p_\alpha$, we can neglect the first term and obtain

$$\frac{1}{2d}\frac{d}{dt}\int_{\mathscr{O}}|Y_\alpha(t)|^{2d}d\xi \; + \; (2d-1)\int_{\mathscr{O}}|Y_\alpha(t)|^{2d-2}|\nabla_\xi Y_\alpha(t)|^2 d\xi$$

$$\le \int_{\mathscr{O}}p_\alpha(W_A(t))Y_\alpha(t)^{2d-1}d\xi.$$

Moreover, by the Poincaré inequality (see (4.16)), it follows that

$$\int_{\mathscr{O}}|Y_\alpha(t)|^{2d-2}|\nabla_\xi Y_\alpha(t)|^2 d\xi = d^{-2}\int_{\mathscr{O}}|\nabla_\xi(Y_\alpha^d(t))|^2 d\xi \ge \frac{\pi^2}{d^2}\int_{\mathscr{O}}|Y_\alpha(t)|^{2d}d\xi.$$

Consequently,

$$\frac{d}{dt}\int_{\mathscr{O}}|Y_\alpha(t)|^{2d}d\xi \le -\frac{2\pi^2}{d}\int_{\mathscr{O}}|Y_\alpha(t)|^{2d}d\xi + 2d\int_{\mathscr{O}}p_\alpha(W_A(t))Y_\alpha(t)^{2d-1}d\xi$$

and there exists a constant $a_1 > 0$ such that

$$\frac{d}{dt}\int_{\mathscr{O}}|Y_\alpha(t)|^{2d}d\xi \le -\frac{\pi^2}{d}\int_{\mathscr{O}}|Y_\alpha(t)|^{2d}d\xi + a_1\int_{\mathscr{O}}(1+|W_A(t)|^{2d^2})d\xi.$$

Consequently,

$$|Y_\alpha(t)|^{2d}_{L^{2d}(\mathscr{O})}d\xi \le e^{-\frac{\pi^2}{d}t}|x|^{2d}_{L^{2d}(\mathscr{O})} + a_1\int_0^t e^{-\frac{\pi^2}{d}(t-s)}\int_{\mathscr{O}}(1+|W_A(s)|^{2d^2})d\xi,$$

and, for a constant $a_2 > 0$,

$$|Y_\alpha(t)|^{2d}_{L^{2d}(\mathscr{O})}d\xi \le e^{-\frac{\pi^2}{d}t}|x|^{2d}_{L^{2d}(\mathscr{O})} + a_2\left(1+\sup_{(s,\xi)\in[0,T]\times\mathscr{O}}|W_A(s,\xi)|^{2d^2}\right).$$

By Theorem 4.8 there exists $a_3 > 0$ such that

$$\mathbb{E}\left(|Y_\alpha(t)|^{2d}_{L^{2d}(\mathscr{O})}\right) \le e^{-\frac{\pi^2}{d}t}|x|^{2d}_{L^{2d}(\mathscr{O})} + a_3,$$

and so, there exists $a_4 > 0$ such that

$$\mathbb{E}\left(|X_\alpha(t,x)|^{2d}_{L^{2d}(\mathscr{O})}\right) \le e^{-\frac{\pi^2}{d}t}|x|^{2d}_{L^{2d}(\mathscr{O})} + a_4, \quad t \ge 0.$$

Letting $\alpha \to 0$ yields

$$\mathbb{E}\left(|X(t,x)|^{2d}_{L^{2d}(\mathscr{O})}\right) \le e^{-\frac{\pi^2}{d}t}|x|^{2d}_{L^{2d}(\mathscr{O})} + a_4, \quad t \ge 0. \qquad (4.30)$$

Denote by $\nu_{t,x}$ the law of $X(t,x)$. Then by (4.30) it follows that for any $\beta > 0$,

$$\int_H \frac{|y|^{2d}_{L^{2d}(\mathcal{O})}}{1 + \beta |y|^{2d}_{L^{2d}(\mathcal{O})}} \, \nu_{t,x}(dy) \leq \int_H |y|^{2d}_{L^{2d}(\mathcal{O})} \, \nu_{t,x}(dy)$$

$$= \mathbb{E}\left(|X(t,x)|^{2d}_{L^{2d}(\mathcal{O})}\right) \leq e^{-\frac{\kappa^2}{d} t} |x|^{2d}_{L^{2d}(\mathcal{O})} + a_4, \quad x \in H, \, t \geq 0.$$

Consequently, letting $t$ tend to $\infty$ and taking into account that $\nu_{t,x}$ is weakly convergent to $\nu$ as $t \to \infty$ by Theorem 4.16, we find that

$$\int_H \int_H \frac{|y|^{2d}_{L^{2d}(\mathcal{O})}}{1 + \beta |y|^{2d}_{L^{2d}(\mathcal{O})}} \, \nu_{t,x}(dy) \leq a_4,$$

which, letting $\beta$ tend to 0, yields the conclusion.      □

By Proposition 4.20 we obtain that

**Corollary 4.21.** *We have*

$$\int_H |F(x)|^2 \nu(dx) < \infty.$$

The corollary implies that $K_0 \varphi \in L^2(H, \nu)$ for all $\varphi \in \mathscr{E}_A(H)$ as required.

Now we can show that $K\varphi = K_0\varphi$ for all $\varphi \in \mathscr{E}_A(H)$. For this we need the following Itô's formula.

**Proposition 4.22.** *For any $\varphi \in \mathscr{E}_A(H)$ we have*

$$\mathbb{E}\left[\varphi(X(t,x))\right] = \varphi(x) + \mathbb{E}\left[\int_0^t K_0\varphi(X(s,x))ds\right], \quad t \geq 0, \, x \in H. \tag{4.31}$$

*Moreover, $\varphi \in D(K_2)$ and $K_2\varphi = K_0\varphi$.*

*Proof.* We omit the proof of (4.31), since it is similar to that of Proposition 3.19. It remains to prove that each $\varphi \in D(K_0)$ belongs to $D(K_2)$ and $K_2\varphi = K_0\varphi$. Since by (4.31) it follows that

$$\lim_{h \to 0} \frac{1}{h} \left(P_h\varphi(x) - \varphi(x)\right) = N_0\varphi(x)$$

pointwise, it is enough to show that

$$\frac{1}{h}\left(P_h\varphi - \varphi\right), \quad h \in (0, 1],$$

is equibounded in $L^2(H, \nu)$.

We have in fact, in view of (4.29) and (4.31),

$$|P_h\varphi(x) - \varphi(x)| \leq \int_0^h \mathbb{E}[a + b|X(s,x)| + \|\varphi\|_1 |F(X(s,x))|]ds, \quad x \in H.$$

By the Hölder inequality we find that

$$|P_h\varphi(x) - \varphi(x)|^2 \leq h \int_0^h [\mathbb{E}(a + b|X(s,x)| + \|\varphi\|_1 |F(X(s,x))|)]^2 ds$$

$$\leq 2h \int_0^h \mathbb{E}[(a + b|X(s,x)|^2)ds + 2h\|\varphi\|_1^2 \int_0^h \mathbb{E}[|F(X(s,x))|^2]ds$$

$$= 2h \int_0^h P_s(a + b|\cdot|^2)(x)ds + 2\|\varphi\|_1^2 h \int_0^h P_s(|F(\cdot)|)(x)ds.$$

Integrating with respect to $\nu$ over $H$, and taking into account the invariance of $\nu$, yields

$$\|P_h\varphi - \varphi\|_{L^2(H,\nu)}^2 \leq 2h^2 \int_H [(a + b|x|^2) + \|\varphi\|_1^2 |F(x)|^2]\nu(dx) < +\infty,$$

thanks to Corollary 4.21. Consequently, $\frac{1}{h}(P_h\varphi - \varphi)$ is equibounded in $L^2(H,\nu)$ as claimed.                                                                                      $\square$

**Theorem 4.23.** *Assume that Hypothesis 4.4 holds with $\lambda = 0$. Then $K_2$ is the closure of $K_0$ in $L^2(H,\nu)$.*

*Proof.* By Proposition 4.22 we know that $K_2$ extends $K_0$. Since $K_2$ is dissipative, so is $K_0$. Consequently, $K_0$ is closable. Let us denote by $\overline{K_0}$ its closure. We have to show that $K_2 = \overline{K_0}$.

Let $\lambda > 0$ and $f \in \mathscr{E}_A(H)$. Consider the approximating equation

$$\lambda\varphi_\alpha - L\varphi_\alpha - \langle F_\alpha, D\varphi_\alpha \rangle = f, \quad \alpha > 0, \tag{4.32}$$

where $F_\alpha$ are the Yosida approximations of $F$. By Step 1 in the proof of Theorem 3.21 equation (4.32) has a unique solution $\varphi_\alpha \in D(L, C_{b,1}^1(H)) \cap C_b^1(H)$ given by

$$\varphi_\alpha(x) = \int_0^{+\infty} e^{-\lambda t}\mathbb{E}[f(X_\alpha(t,x))]dt, \quad x \in H,$$

where $X_\alpha(\cdot, x)$ is the solution to problem (4.9). Moreover, for all $h \in H$ we have

$$\langle D\varphi_\alpha(x), h \rangle = \int_0^{+\infty} e^{-\lambda t}\mathbb{E}[\langle Df(X_\alpha(t,x)), \eta_\alpha^h(t,x)h \rangle]dt, \tag{4.33}$$

where for any $h \in H$, $\eta_\alpha^h(t, x) = DX_\alpha(t, x) \cdot h$. By (4.17) it follows that

$$|\eta_\alpha^h(t, x)| \leq e^{-\pi^2 t}|h|, \quad t \geq 0, \ h \in H.$$

Consequently by (4.33) we obtain

$$|D\varphi_\alpha(x)| \leq \frac{1}{\lambda}\|f\|_1, \quad x \in H. \tag{4.34}$$

We claim that

$$D(L, C_{b,1}(H)) \cap C_b^1(H) \subset D(\overline{K_0}). \tag{4.35}$$

In fact let $\varphi \in D(L, C_{b,1}(H)) \cap C_b^1(H)$. Then, by Proposition 2.72, there exists a three-index sequence $\{\varphi_{n_1,n_2,n_3}\} \subset \mathscr{E}_A(H)$ such that

$$\lim_{n_1 \to \infty} \lim_{n_2 \to \infty} \lim_{n_3 \to \infty} \varphi_{n_1,n_2,n_3}(x) = \varphi(x), \quad x \in H,$$

$$\lim_{n_1 \to \infty} \lim_{n_2 \to \infty} \lim_{n_3 \to \infty} D\varphi_{n_1,n_2,n_3}(x) = D\varphi(x), \quad x \in H,$$

$$\lim_{n_1 \to \infty} \lim_{n_2 \to \infty} \lim_{n_3 \to \infty} L\varphi_{n_1,n_2,n_3}(x) = L\varphi(x), \quad x \in H.$$

Then by the dominated convergence theorem (and Corollary 4.21) it follows that

$$\lim_{n_1 \to \infty} \lim_{n_2 \to \infty} \lim_{n_3 \to \infty} K_0\varphi_{n_1,n_2,n_3}(x) = K_2\varphi(x) = L\varphi(x) + \langle F(x), D\varphi(x)\rangle, \quad x \in H.$$

So, (4.35) is proved. Now we can write (4.32) as

$$\lambda\varphi_\alpha - \overline{K_0}\varphi_\alpha = f + \langle F_\alpha - F, D\varphi_\alpha\rangle.$$

We claim that

$$\lim_{\alpha \to 0} \langle F_\alpha - F, D\varphi_\alpha\rangle = 0 \quad \text{in } L^2(H, \nu).$$

We have in fact, in view of (4.34),

$$\int_H |\langle F_\alpha(x) - F(x), D\varphi_\alpha\rangle|^2 \nu(dx) \leq \frac{1}{\lambda^2}\|f\|_1^2 \int_H |F_\alpha(x) - F(x)|^2 \nu(dx).$$

Clearly,

$$\lim_{\alpha \to 0} |F_\alpha(x) - F(x)|^2 = 0 \quad \nu\text{–a.e.}$$

Moreover

$$|F_\alpha(x) - F(x)|^2 \leq 2|F(x)|^2, \quad x \in H.$$

Therefore, the claim follows from the dominated convergence theorem, since $\int_H |F(x)|^2 \nu(dx)$ is finite in view of Proposition 4.20.

In conclusion we have proved that the closure of the range of $\lambda - \overline{K_0}$ includes $\mathscr{E}_A(H)$ which is dense in $L^2(H, \nu)$ and the theorem follows from the Lumer–Phillips theorem 3.20. $\qquad \square$

## 4.7 The integration by parts formula and its consequences

In this section we shall assume that Hypothesis 4.4 holds with $\lambda = 0$. We still denote by $\nu$ the unique invariant measure for $P_t$.

The following two propositions can be proved as Propositions 3.22 and 3.23 respectively and so, we shall omit the corresponding proofs for brevity.

**Proposition 4.24.** *The operator* $D_C \colon \mathscr{E}_A(H) \to C_b(H;H)$, $\varphi \mapsto C^{1/2}D\varphi$ [7] *is uniquely extendible to a bounded operator, still denoted by* $D_C$, *from* $D(K_2)$ *(endowed with the graph norm) into* $L^2(H,\nu;H)$. *Moreover the following identity holds:*

$$\int_H K_2\varphi\, \varphi\, d\nu = -\frac{1}{2} \int_H |D_C\varphi|^2 d\nu, \qquad \varphi \in D(K_2). \tag{4.36}$$

**Proposition 4.25.** *Let* $\varphi \in L^2(H,\nu)$ *and* $t \geq 0$. *Then, for any* $T > 0$, *the linear operator*

$$D_C \colon D(K_2) \to L^2(0,T;L^2(H,\nu;H)), \quad \varphi \mapsto D_C P_t\varphi,$$

*is uniquely extendible to a bounded operator, still denoted by* $D_C$, *from* $L^2(H,\nu)$ *into* $L^2(0,T;L^2(H,\nu;H))$. *Moreover the following identity holds:*

$$\int_H (P_t\varphi)^2\, d\nu + \int_0^t ds \int_H |D_C P_s\varphi|^2 d\nu = \int_H \varphi^2\, d\nu. \tag{4.37}$$

### 4.7.1 The Sobolev space $W^{1,2}(H,\nu)$

Here and in the next subsection we shall assume in addition that $\gamma = 0$ (so that $n = 1$), $C = 1$ and follow [37].

The following result can be proved as Theorem 3.25. So, we shall omit the proof.

**Theorem 4.26.** *The linear operator*

$$D \colon \mathscr{E}_A(H) \to L^2(H,\nu), \quad \varphi \mapsto D\varphi,$$

*is closable. Moreover, if* $\varphi$ *belongs to the domain* $\overline{D}$ *of the closure of* $D$ *and* $\overline{D}\varphi = 0$, *we have that* $\overline{D}P_t\varphi = 0$ *for any* $t > 0$.

We shall denote by $W^{1,2}(H,\nu)$ the domain of $\overline{D}$ and, if there is no possibility of confusion, we shall set $\overline{D} = D$.

The following result can be proved as Proposition 3.26.

**Proposition 4.27.** *We have* $D(K_2) \subset W^{1,2}(H,\nu)$ *with continuous embedding. Moreover, the following identity holds:*

$$\int_H K_2\varphi\, \varphi\, d\nu = -\frac{1}{2} \int_H |D\varphi|^2 d\nu, \qquad \varphi \in D(K_2).$$

---

[7] Recall that $C = (-A)^{-\gamma}$.

## 4.7.2  Poincaré and log-Sobolev inequalities, spectral gap

To prove Poincaré and log-Sobolev inequalities we shall use Propositions 3.28 and 3.29 with $F_\alpha$ replacing $F$ and then we let $\alpha \to 0$. For this we need the following result.

**Proposition 4.28.** *The sequence $\{\nu_\alpha\}$ is weakly convergent to $\nu$ as $\alpha \to 0$.*

*Proof.* It is enough to show that the sequence of measures $\{\nu_\alpha\}$ is tight, because this will imply that a subsequence $\{\nu_{\alpha_k}\}$ is weakly convergent to the invariant measure $\nu$. By the uniqueness of $\nu$, the whole sequence $\{\nu_\alpha\}$ will be weakly convergent to $\nu$ as $\alpha \to 0$.

We first notice that, setting $Y_\alpha(t) = X_\alpha(t) - W_A(t)$, (4.9) reduces to

$$\begin{cases} Y'_\alpha(t) = AY_\alpha(t) - p_\alpha(Y_\alpha(t) + W_A(t)), & t \in [0,T], \\ Y_\alpha(0) = x. \end{cases}$$

Multiplying scalarly both sides of the first equation by $Y_\alpha(t)$ yields

$$\frac{1}{2}\frac{d}{dt}\int_{\mathcal{O}} |Y_\alpha(t)|^2 d\xi + \int_{\mathcal{O}} |\nabla_\xi Y_\alpha(t)|^2 d\xi$$

$$= \int_{\mathcal{O}} [p_\alpha(Y_\alpha(t) + W_A(t)) - p_\alpha(W_A(t))]d\xi + \int_{\mathcal{O}} p_\alpha(W_A(t))d\xi.$$

Taking into account the monotonicity of $p_\alpha$, we can neglect the second term and obtain

$$\frac{1}{2}\frac{d}{dt}\int_{\mathcal{O}} |Y_\alpha(t)|^2 d\xi + \int_{\mathcal{O}} |D_\xi Y_\alpha(t)|^2 d\xi \leq \int_{\mathcal{O}} p_\alpha(W_A(t))d\xi.$$

It follows that

$$\frac{d}{dt}\int_{\mathcal{O}} |Y_\alpha(t)|^2 d\xi + \int_{\mathcal{O}} |D_\xi Y_\alpha(t)|^2 d\xi \leq -\pi^2 \int_{\mathcal{O}} |Y_\alpha(t)|^2 d\xi + \int_{\mathcal{O}} p_\alpha(W_A(t))d\xi.$$

Consequently,

$$\int_{\mathcal{O}} |Y_\alpha(t)|^2 d\xi + \int_0^t ds \int_{\mathcal{O}} |D_\xi Y_\alpha(s)|^2 d\xi \leq e^{-\pi^2 t}|x|^2 + \int_0^t e^{-\pi^2(t-s)}\int_{\mathcal{O}} p_\alpha(W_A(s))d\xi.$$

Taking expectation, we find, after some standard manipulation,

$$\frac{1}{t}\,\mathbb{E}\int_0^t ds \int_{\mathcal{O}} |D_\xi Y_\alpha(s)|^2 d\xi \leq c_1,$$

where $c_1$ is a suitable constant.

We notice now that we cannot get an estimate for $X_\alpha$ simply by replacing $Y_\alpha$ by $X - W_A$ because

$$\mathbb{E} \int_{\mathscr{O}} |D_\xi W_A(s)|^2 d\xi = +\infty.$$

However, it is easy to check that

$$\mathbb{E} \int_{\mathscr{O}} |(-A)^\beta W_A(s)|^2 d\xi < +\infty,$$

provided $\beta \in (0, 1/2)$.

By interpolation it follows that there is $\beta \in (0, 1/2)$ such that

$$\mathbb{E} \int_{\mathscr{O}} |(-A)^\beta X_\alpha(s)|^2 d\xi < c_\beta.$$

Now, integrating with respect to $\nu_\alpha$ yields

$$\int_H |(-A)^\beta x|^2 \nu_\alpha(dx),$$

which proves tightness because the operator $(-A)^{-\beta}$ is compact in $H$.                  □

We are now in a position to prove the announced inequalities. We start with the Poincaré inequality.

**Proposition 4.29.** *For any $\varphi \in W^{1,2}(H, \nu)$ we have*

$$\int_H |\varphi - \overline{\varphi}|^2 d\nu \le \frac{1}{2\pi^2} \int_H |D\varphi|^2 d\nu, \qquad (4.38)$$

*where*

$$\overline{\varphi} = \int_H \varphi d\nu.$$

*Proof.* In view of Theorem 4.26, it is enough to show (4.38) for $\varphi \in \mathscr{E}_A(H)$. Let $\alpha > 0$ and let $\nu_\alpha$ be the invariant measure of the approximating semigroup $P_t^\alpha$. By Proposition 4.28 we have that $\nu_\alpha \rightharpoonup \nu$ weakly as $\alpha \to 0$.

Since by Proposition 3.28 we have

$$\int_H |\varphi - \overline{\varphi}|^2 d\nu_\alpha \le \frac{1}{2\pi^2} \int_H |D\varphi|^2 d\nu_\alpha,$$

the conclusion follows letting $\alpha$ tend to 0.                  □

The proof of the following result is completely analogous to that of Proposition 3.29.

**Proposition 4.30.** *We have*

$$\sigma(K_2)\backslash\{0\} \subset \{\lambda \in \mathbb{C} : \Re\, \lambda \le -\pi^2\}, \tag{4.39}$$

*and*

$$\int_{\mathbb{R}} |P_t\varphi - \overline{\varphi}|^2 d\nu \le e^{-2\pi^2 t} \int_{\mathbb{R}} |\varphi|^2 d\nu, \quad \varphi \in L^2(H,\nu).$$

Let us finally prove the log-Sobolev inequality.

**Proposition 4.31.** *For any $\varphi \in W^{1,2}(H,\nu)$ we have*

$$\int_H \varphi^2 \log(\varphi^2) d\nu \le \frac{1}{\pi^2} \int_H |D\varphi|^2 d\nu + |\varphi|^2_{L^2(H,\nu)} \log(|\varphi|^2_{L^2(H,\nu)}). \tag{4.40}$$

*Proof.* Again, in view of Theorem 4.26, it is enough to show (4.38) for $\varphi \in \mathscr{E}_A(H)$. Let $\alpha > 0$ and let $\nu_\alpha$ be the invariant measure of the approximating semigroup $P_t^\alpha$.

By Proposition 3.30 we have

$$\int_H \varphi^2 \log(\varphi^2) d\nu_\alpha \le \frac{1}{\pi^2} \int_H |D\varphi|^2 d\nu_\alpha + |\varphi|^2_{L^2(H,\nu_\alpha)} \log(|\varphi|^2_{L^2(H,\nu_\alpha)}),$$

and the conclusion follows letting $\alpha$ tend to 0. $\square$

**Remark 4.32.** Proceeding as in the proof of Theorem 2.56, we can prove that if $P_t$ is symmetric, then $P_t$ is hypercontractive.

## 4.8 Comparison of ν with a Gaussian measure

In this section we still assume Hypothesis 4.4 with $\lambda = 0$ and $\gamma = 0$. We follow [35]. In this case, the Ornstein–Uhlenbeck operator $L$ in $C_b(H)$ has a unique invariant measure $\mu = N_{Q_\infty}$ where $Q_\infty = -\frac{1}{2} A^{-1}$. Moreover, $P_t$ has a unique invariant measure $\nu$. We want here to show that $\nu$ is absolutely continuous with respect to $\mu$.

Let us consider the Kolmogorov operators $K, K_\alpha$, $\alpha > 0$, defined as

$$K\varphi(x) = L\varphi(x) + \langle F(x), D\varphi(x)\rangle, \quad \varphi \in D(L, C_b(H)),$$

and

$$K_\alpha\varphi(x) = L\varphi(x) + \langle F_\alpha(x), D\varphi(x)\rangle, \quad \varphi \in D(L, C_b(H)).$$

**Lemma 4.33.** *For any $\lambda > 0$, $\alpha > 0$ and any $f \in B_b(H)$ the following identity holds:*

$$(\lambda - L)^{-1}f = (\lambda - K_\alpha)^{-1}f - (\lambda - K_\alpha)^{-1}\left[\langle F_\alpha, D(\lambda - L)^{-1}f\rangle\right]. \tag{4.41}$$

*Proof.* Notice that we have proved in §3.7.1 this identity when $F$ is Lipschitz and bounded. To show (4.41) in the present case we shall approximate $F_\alpha$ (which is Lipschitz but not bounded) by Lipschitz bounded functions.

Set, for any $\varepsilon > 0$,

$$F_{\alpha,\varepsilon}(x) = \frac{F_\alpha(x)}{1 + \varepsilon|x|}, \quad x \in H.$$

Then $F_{\alpha,\varepsilon}$ are Lipschitz continuous, uniformly in $\varepsilon$, and bounded. We denote by $X_{\alpha,\varepsilon}(t,x)$ the mild solution of the stochastic differential equation

$$dX = (AX + F_{\alpha,\varepsilon}(X))dt + \sqrt{C}\,dW_t, \quad X(0) = x,$$

by $P_t^{\alpha,\varepsilon}$ the corresponding transition semigroup

$$P_t^{\alpha,\varepsilon}\varphi(x) = \mathbb{E}\left[\varphi(X_{\alpha,\varepsilon}(t,x))\right], \quad \varphi \in B_b(H),$$

and by $K_{\alpha,\varepsilon}$ its infinitesimal generator defined as before. Now, let $\lambda > 0$, $f \in B_b(H)$ and consider the equation

$$\lambda\varphi_{n,\varepsilon} - L\varphi_{n,\varepsilon} - \langle F_{n,\varepsilon}, D\varphi_{n,\varepsilon}\rangle = f. \tag{4.42}$$

Setting $\lambda\varphi_{n,\varepsilon} - L\varphi_{n,\varepsilon} = \psi_{n,\varepsilon}$, (4.42) becomes

$$\psi_{n,\varepsilon} - T_\lambda^{n,\varepsilon}\psi_{n,\varepsilon} = f, \tag{4.43}$$

where

$$T_\lambda^{n,\varepsilon}\psi = \langle F_{n,\varepsilon}, DR(\lambda - L)^{-1}\psi\rangle, \quad \psi \in B_b(H).$$

But, in view of Proposition 2.29, we have

$$\|T_\lambda^{n,\varepsilon}\psi\|_0 \le \lambda_{\alpha,\varepsilon} := \sqrt{\frac{\pi}{\lambda}}\,\|F_{n,\varepsilon}\|_0\,\|\psi\|_0, \quad \psi \in B_b(H).$$

Consequently, for $\lambda > \pi\|\|F_{n,\varepsilon}\|_0$, equation (4.42) can be uniquely solved by a standard fixed point argument. In conclusion, $(\lambda_{\alpha,\varepsilon}, \infty)$ belongs to the resolvent set of $K_{\alpha,\varepsilon}$ and we have

$$(\lambda - K_{\alpha,\varepsilon})^{-1} = (\lambda - L)^{-1}(1 - T_\lambda^{\alpha,\varepsilon})^{-1}, \quad \text{for } \lambda > \lambda_{\alpha,\varepsilon}.$$

It follows that, for $\lambda > \lambda_{\alpha,\varepsilon}$,

$$(\lambda - L)^{-1}f = (\lambda - K_{\alpha,\varepsilon})^{-1}f - (\lambda - K_{\alpha,\varepsilon})^{-1}\left[\langle F_{\alpha,\varepsilon}, D(\lambda - L)^{-1}f\rangle\right]. \tag{4.44}$$

Now, by analytic continuation (4.44) holds for any $\lambda > 0$. Finally the conclusion follows by letting $\varepsilon$ tend to 0. □

**Theorem 4.34.** *Assume that Hypothesis 4.4 with $\lambda = 0$ and $\gamma = 0$ holds. Let $\mu$ and $\nu$ be the invariant measures of $R_t$ and $P_t$ respectively. Then for any $f \in B_b(H)$ we have*

$$\int_H f d\mu = \int_H f d\nu + \int_H \langle F, DL^{-1}f \rangle d\nu. \tag{4.45}$$

*Moreover $\nu$ is absolutely continuous with respect to $\mu$ and $D \log \rho \in L^2(H,\nu)$ where $\rho = \frac{d\nu}{d\mu}$.*

*Proof.* By Lemma 4.33 we have, for any $\alpha > 0$, $r > 0$,

$$\int_H f d\mu = \int_H f d\nu_\alpha + \int_H \langle F, DL^{-1}f \rangle d\nu_\alpha. \tag{4.46}$$

We know by Proposition 4.28 that $\nu_\alpha \rightharpoonup \nu$ as $\alpha \to 0$. Consequently

$$\lim_{\alpha \to 0} \int_H f d\nu_\alpha = \int_H f d\nu.$$

Moreover, taking into account Corollary 4.21, it is not difficult to see that

$$\lim_{\alpha \to 0} \int_H \langle F, DL^{-1}f \rangle d\nu_\alpha = \int_H \langle F, DL^{-1}f \rangle d\nu. \tag{4.47}$$

Therefore, letting $\alpha \to 0$ in (4.46) yields (4.45).

We are now ready to prove the absolute continuity of $\nu$ with respect to $\mu$. Let $\Gamma \subset H$ be a Borel set such that $\mu(\Gamma) = 0$. Then we have

$$R_t 1_\Gamma(x) = N_{e^{tA}x, Q_t}(\Gamma) = 0, \text{ for all } t > 0 \text{ and } x \in H.$$

This follows because $R_t$ is strong Feller and the measure $N_{e^{tA}x, Q_t}$ is absolutely continuous with respect to $\mu$. Consequently, $D(\lambda - L)^{-1}1_\Gamma(x) = 0$ for all $x \in H$, $\lambda > 0$. Thus, by (4.44) it follows that $\nu(\Gamma) = \mu(\Gamma) = 0$. $\qquad\square$

## 4.9 Compactness of the embedding $W^{1,2}(H,\nu) \subset L^2(H,\nu)$

Here we shall assume for simplicity that Hypothesis 4.4 with $\lambda = 0$ and $\gamma = 0$ holds. We follow [37] where a more general result is proved.

**Theorem 4.35.** *Assume that Hypothesis 4.4 with $\lambda = 0$ and $\gamma = 0$ holds. Let $\nu$ and $\mu$ be the invariant measures of $P_t$ and $R_t$ respectively. Assume in addition that there exists $\varepsilon \in [0,1]$ such that*

$$\int_H |D \log \rho|^{2+\varepsilon} d\nu < +\infty,$$

*where $\rho = \frac{d\nu}{d\mu}$. Then the embedding $W^{1,2}(H,\nu) \subset L^2(H,\nu)$ is compact.*

*Proof.* Let $\{\varphi_n\} \subset W^{1,2}(H, \nu)$ be such that

$$\int_H |\varphi_n|^2 d\nu + \int_H |D\varphi_n|^2 d\nu \leq 1, \quad n \in \mathbb{N}.$$

We have to show that there exists a subsequence $\{\varphi_{n_k}\}$ which is convergent in $L^2(H, \nu)$. Notice that, in view of the log-Sobolev inequality (4.40), the sequence $\{\varphi_n^2\}$ is uniformly integrable. Thus it is enough to find a subsequence $\{\varphi_{n_k}\}$ that is pointwise convergent, since then, by the Vitali theorem, one can conclude that $\{\varphi_{n_k}\}$ is convergent in $L^2(H, \nu)$.

To construct a pointwise convergent subsequence, we proceed as follows. First for any $R > 1$ we set

$$G_R = \left\{ x \in H : \rho(x) \geq \frac{1}{R} \right\},$$

and notice that $\lim_{R \to \infty} \nu(G_R) = 1$. Then we consider a function $\theta : [0, +\infty) \to [0, +\infty)$ of class $C^\infty$ such that

$$\theta(r) \begin{cases} = 1 & \text{if } r \in [0, 1], \\ = 0 & \text{if } r \geq 2, \\ \in [0, 1] & \text{if } r \in [1, 2], \end{cases}$$

and set

$$\varphi_{n,R}(x) = \theta\left( \frac{-2\log\rho(x)}{\log R} \right) \varphi_n(x), \quad x \in H,$$

so that

$$\varphi_{n,R}(x) \begin{cases} \leq |\varphi_n(x)|, & x \in H, \\ = 0 & \text{if } \rho(x) < \frac{1}{R}, \\ = \varphi_n(x) & \text{if } \rho(x) \geq \frac{1}{\sqrt{R}}. \end{cases}$$

Finally, we prove that there exists $C_R > 0$ and $\alpha \in (0, 1)$ such that

$$\int_H |\varphi_{n,R}|^2 d\mu + \int_H |D\varphi_{n,R}|^{1+\alpha} d\mu \leq C_R. \tag{4.48}$$

Once the claim is proved, since the embedding $W^{1,1+\alpha}(H, \mu) \subset L^{1+\alpha}(H, \mu)$ is compact see [21], we can construct a subsequence $\{\varphi_{n_k,R}\}$ which is convergent in $L^{1+\alpha}(H, \nu)$ and then another subsequence which is pointwise convergent. Now, by a standard diagonal procedure we can find a subsequence $\{\varphi_{n_k}\}$ pointwise convergent as required.

It remains to show (4.48). We have in fact

$$\int_H |\varphi_{n,R}|^2 d\mu = \int_{\{\rho \geq \frac{1}{R}\}} |\varphi_{n,R}|^2 \frac{d\nu}{\rho} \leq R \int_H |\varphi_n|^2 d\mu \leq R. \tag{4.49}$$

Moreover

$$D\varphi_{n,R}(x) = -2\theta'\left(\frac{-2\log\rho(x)}{\log R}\right)\left(\frac{D\log\rho(x)}{\log R}\right)\varphi_n(x)$$

$$+ \theta\left(\frac{-2\log\rho(x)}{\log R}\right)D\varphi_n(x) := F_{n,R}(x) + H_{n,R}(x)$$

and

$$\int_H |H_{n,R}|^2 d\mu \le \int_{\{\rho \ge \frac{1}{R}\}} |D\varphi_{n,R}|^2 \frac{d\nu}{\rho} \le R \int_H |D\varphi_n|^2 d\mu \le R. \tag{4.50}$$

Also, for any $\alpha \in (0,1)$ we have, using the Hölder inequality,

$$\int_H |F_{n,R}|^{1+\alpha} d\mu \le \left(\frac{2}{\log R}\right)^{1+\alpha} R \int_H |D\log\rho|^{1+\alpha}|\varphi_n|^{1+\alpha} d\mu$$

$$\le \left(\frac{2}{\log R}\right)^{1+\alpha} R\left(\int_H |D\log\rho|^{\frac{2(1+\alpha)}{1-\alpha}} d\nu\right)^{\frac{1-\alpha}{2}}\left(\int_H |\varphi_n|^2 d\nu\right)^{\frac{1+\alpha}{2}}. \tag{4.51}$$

Choosing $\alpha = \frac{\varepsilon}{4+\varepsilon}$ we have $\frac{2(1+\alpha)}{1-\alpha} = 2 + \varepsilon$. Then by (4.50) and (4.51), (4.48) follows. □

## 4.10 Gradient systems

Here we assume that Hypothesis 4.4 holds with $\gamma = 0$ and $\lambda = 0$. So, equation (4.4) reads as

$$\begin{cases} dX(t) = [AX(t) - p(X(t))]dt + dW(t), \\ \\ X(0) = x, \end{cases} \tag{4.52}$$

where $p$ is a polynomial such that $p'(\xi) \ge 0$. We know that the transition semigroup $P_t$ corresponding to (4.52) has a unique invariant measure $\nu$ and that the corresponding Ornstein–Uhlenbeck semigroup $R_t$ is symmetric and possesses a unique invariant measure $\mu = N_{Q_\infty}$. Moreover, by Theorem 4.34 $\nu << \mu$ and if $\rho = \frac{d\nu}{d\mu}$ we have $D\log\rho \in L^2(H,\nu)$.

We want to show now that

$$\nu(dx) = \frac{e^{-2U(x)}}{\int_H e^{-2U(y)}}\mu(dx) := Z^{-1}e^{-2U(x)}\mu(dx), \tag{4.53}$$

where

$$U(x) = \int_0^1 p(x(\xi))d\xi, \quad x \in L^{2d}(0,1). \tag{4.54}$$

To this purpose let us first consider the approximating problem

$$
\begin{cases}
dX_\alpha(t) = [AX_\alpha(t) - p_\alpha(X(t))]dt + dW(t), \\
\\
X(0) = x,
\end{cases}
\tag{4.55}
$$

where $\alpha > 0$ and $p_\alpha$ is the Yosida approximation of $p$. We know that the transition semigroup $P_t^\alpha$ corresponding to (4.55) has a unique invariant measure $\nu_\alpha$ and that, $\nu_\alpha << \mu$.

**Lemma 4.36.** *We have*

$$
\nu_\alpha(dx) = \frac{e^{-2U_\alpha(x)}}{\int_H e^{-2U(y)}}\, \mu(dx) := Z^{-\alpha} e^{-2U(x)} \mu(dx),
\tag{4.56}
$$

*where*

$$
U_\alpha(x) = \int_0^1 p_\alpha(x(\xi))d\xi, \quad x \in L^{2d}(0,1).
\tag{4.57}
$$

*Moreover, $P_t^\alpha$ is symmetric.*

*Proof.* By the uniqueness of $\nu_\alpha$, to show (4.56) it is enough to show that the measure $\zeta_\alpha$, where

$$
\zeta_\alpha(dx) = Z^{-\alpha} e^{-2U(x)} \mu(dx),
$$

is invariant for $P_t^\alpha$. Let $K_2^\alpha$ be the infinitesimal generator of $P_t^\alpha$. Then for any $\varphi \in C_b^1(H)$ we have, by a simple computation,

$$
\int_H K_2^\alpha \varphi d\zeta_\alpha = Z^{-\alpha} \int_H L_2 e^{-2U(x)} d\mu + \int_H \langle p_\alpha(x), D\varphi \rangle d\mu = 0.
$$

Therefore, (4.56) holds. Symmetry of $K_2^\alpha$ follows from a simple computation.   $\square$

We can now prove

**Proposition 4.37.** *$P_t$ is symmetric and (4.53) is fulfilled.*

*Proof.* Thanks to Lemma 4.36 it is enough to show that $\{\nu_\alpha\}$ is tight. By this follows from Proposition 4.28.   $\square$

**Remark 4.38.** Let $\rho = \frac{d\nu}{d\mu}$. Since

$$
D \log \rho(x) = p(x),
$$

it follows that the assumptions of Theorem 4.35 are fulfilled. So the embedding

$$
W^{1,2}(H,\nu) \subset L^2(H,\nu),
$$

is compact.

# Chapter 5

# The Stochastic Burgers Equation

## 5.1 Introduction and preliminaries

We are here concerned with the *Burgers* equation perturbed by noise. For the sake of simplicity we shall only consider the case when the equation is equipped with periodic boundary conditions. For the case of Dirichlet boundary conditions see [36].

We denote by $H$ the Hilbert space of all $2\pi$-periodic real measurable functions $x$ such that

$$\int_0^{2\pi} |x(\xi)|^2 d\xi < +\infty,$$

endowed with the inner product

$$\langle x, y \rangle_2 = \int_0^{2\pi} x(\xi)y(\xi)d\xi, \quad x, y \in H,$$

and consider the stochastic Burgers equation in $H$,

$$\begin{cases} dX = \left( D_\xi^2 X - X + \dfrac{1}{2} D_\xi(X^2) \right) dt + \sqrt{C}\, dW, \quad \xi \in [0,1], t \geq 0, \\[2mm] X(t, \cdot) \text{ is periodic with period } 2\pi, \quad t \geq 0, \\[2mm] X(0, \xi) = x(\xi), \quad \xi \in [0, 2\pi], \end{cases} \tag{5.1}$$

where $x \in H$, $C \in L_1^+(H)$ and $W$ is a cylindrical Wiener process in a probability space $(\Omega, \mathscr{F}, \mathbb{P})$ in H.

Let us write problem (5.1) in the usual abstract form. For this it is convenient to consider the complexification $H_{\mathbb{C}}$ of $H$ and the complete orthonormal system $\{e_k\}_{k\in\mathbb{Z}}$ in $H_{\mathbb{C}}$ given by

$$e_k(\xi) = \frac{1}{\sqrt{2\pi}}\, e^{ik\xi}, \quad \xi \in [0, 2\pi], \; k \in \mathbb{Z}.$$

Any element $x \in H$ can be represented by its Fourier series

$$x = \sum_{k\in\mathbb{Z}} x_k e_k, \quad x_k = \langle x, e_k \rangle.$$

For any $\sigma \geq 0$ we define

$$\|x\|_\sigma^2 = \sum_{k\in\mathbb{Z}}(1 + k^2)^{\sigma/2}|x_k|^2$$

and set

$$H_\#^\sigma := \{x \in H : \|x\|_\sigma^2 < +\infty\}.$$

$H_\#^\sigma$ is Hilbert space with the norm $\|\cdot\|_\sigma$. When $\sigma = 0$, $H_\#^\sigma$ reduces to $H$ and we set $\|x\|_0 = |x|_2$.

Now, we define a linear operator $A\colon D(A) \subset H \to H$ and a nonlinear operator $b\colon D(b) \subset H \to H$, setting

$$Ax = D_\xi^2 x - x, \quad x \in D(A) = H_\#^2,$$

and

$$b(x) = \frac{1}{2}\, D_\xi(x^2), \quad x \in D(b) = H_\#^1.$$

The operator $b$ enjoys the crucial property

$$\langle b(x), x \rangle = 0 \quad \text{for all } x \in H_\#^1. \tag{5.2}$$

In fact for any $x \in H_\#^1$ we have

$$\langle b(x), x \rangle_2 = \frac{1}{2} \int_0^{2\pi} D_\xi(x^2(\xi))\, x(\xi) d\xi$$

$$= \frac{1}{3} \int_0^{2\pi} D_\xi(x^3(\xi)) d\xi = \frac{1}{3}\left(x^3(2\pi) - x^3(0)\right) = 0.$$

Finally, problem (5.1) can be written as

$$\begin{cases} dX &= (AX + b(X))dt + \sqrt{C}\, dW(t), \\[2mm] X(0) &= x. \end{cases} \tag{5.3}$$

For the sake of simplicity, we choose $C = (-A)^{-\gamma}$ with $\gamma > 1/2$ so that $C$ is of trace class.

We denote by $A_{\mathbb{C}}$ the complexification of $A$. The linear operator $A_{\mathbb{C}}$ is self-adjoint and

$$A_{\mathbb{C}} e_k = -(1 + k^2) e_k, \quad k \in \mathbb{Z}.$$

As before, we consider the cylindrical Wiener process $W(t)$ setting (formally),

$$W(t) = \sum_{k \in \mathbb{Z}} \beta_k(t) e_k, \quad t \geq 0,$$

where $\{\beta_k\}$ is a sequence of mutually independent standard Brownian motions in $(\Omega, \mathscr{F}, \mathbb{P})$. Moreover, we consider the stochastic convolution

$$W_A(t) = \int_0^t e^{(t-s)A}(-A)^{-\gamma/2} dW(s), \quad t \in [0, T].$$

We shall use several properties of $W_A(t)$ proved in §2.2. We recall in particular that $W_A \in C([0, T]; C([0, 2\pi]))$, $\mathbb{P}$-almost surely.

We end this section by giving additional notation and some preliminaries. For any $p \geq 1$ we denote by $L^p$ the space of all $2\pi$-periodic functions $x$ such that

$$|x|_p := \left( \int_0^{2\pi} |x(\xi)|^p d\xi \right)^{1/p} < +\infty,$$

by $L^\infty$ the space of all $2\pi$-periodic essentially bounded functions endowed with the norm

$$|x|_\infty := \sup_{\xi \in [0, 2\pi]} |x(\xi)|$$

and by $C_\#$ the space of all $2\pi$-periodic continuous functions on $\mathbb{R}$.

We have clearly

$$D((-A)^\sigma) = H_\#^{2\sigma},$$

for all $\sigma \in (0, 1)$.

**Lemma 5.1.** *For any $\sigma \in (0, 1)$ there exists $c_\sigma > 0$ such that*

$$|(-A)^\sigma e^{tA} x|_2 \leq c_\sigma t^{-\sigma} |x|_2, \quad t > 0, \ x \in H.$$

*Proof.* We have in fact by the Parseval identity,

$$|(-A)^\sigma e^{tA} x|_2^2 = \sum_{k \in \mathbb{Z}} (1 + k^2)^\sigma e^{-2(1+k^2)t} x_k^2$$

$$\leq \sup_{k \in \mathbb{Z}} \left[ (1 + k^2)^\sigma e^{-2(1+k^2)t} \right] |x|_2^2$$

and the conclusion follows. $\qquad\square$

**Lemma 5.2.** *There exists $\kappa > 0$ such that*

$$\left| D_\xi e^{tA} x \right|_\infty \leq \kappa t^{-\frac{3}{4}} e^{-t} |x|_2, \quad t > 0, \ x \in H. \tag{5.4}$$

*Proof.* For any $t > 0$ we have

$$D_\xi e^{tA} x = \sum_{k \in \mathbb{Z}} i k e^{-(1+k^2)t} x_k e_k, \quad x \in H.$$

Using the Hölder inequality it follows that for any $\xi \in \mathbb{R}$,

$$\left| D_\xi e^{tA} x(\xi) \right| \ \leq \ \frac{1}{\sqrt{2\pi}} \sum_{k \in \mathbb{Z}} k e^{-(1+k^2)t} |x_k|$$

$$\leq \ \frac{1}{\sqrt{2\pi}} \, e^{-t} \left( \sum_{k \in \mathbb{Z}} k^2 e^{-2k^2 t} \right)^{1/2} |x|_2.$$

Since, as easily seen, $\sum_{k \in \mathbb{Z}} k^2 e^{-2\pi^2 k^2 t}$ behaves as $t^{-3/2}$ near 0, the conclusion follows. $\qquad \square$

We recall the Sobolev embedding theorem. For any $\epsilon > 0$ we have

$$H_\#^\epsilon \subset \begin{cases} L^{\frac{2}{1-2\epsilon}} & \text{if } \epsilon \in (0, 1/2), \\[2mm] C_\#(\mathbb{R}) & \text{if } \epsilon > 1/2, \end{cases}$$

with continuous inclusions. In particular $H_\#^{\frac{1}{4}} \subset L^4$.

We shall use the classical interpolatory estimate,

$$\|x\|_\beta \leq \|x\|_\alpha^{\frac{\gamma-\beta}{\gamma-\alpha}} \|x\|_\gamma^{\frac{\beta-\alpha}{\gamma-\alpha}}, \quad \alpha < \beta < \gamma, \tag{5.5}$$

which follows immediately from the definition of the norms in the different spaces, and the following estimate due to Agmon concerning the limit exponent $\frac{1}{2}$.

**Proposition 5.3.** *For any $x \in H_\#^{\frac{1}{2}}$ we have*

$$|x|_\infty \leq |x|_2^{\frac{1}{2}} \|x\|_1^{\frac{1}{2}}. \tag{5.6}$$

*Proof.* Let $\xi \in [0, 2\pi]$ and $r > 0$. Choose $\eta \in [0, 2\pi]$ such that $|\xi - \eta| \geq r$. Integrating the inequality

$$|x(\xi)|^2 \leq 2|x(\xi) - x(\eta)|^2 + 2|x(\eta)|^2,$$

with respect to $\eta$ from $\xi$ to $\xi + r$, we find

$$\int_\xi^{\xi+r} |x(\xi)|^2 d\eta \leq 2 \int_\xi^{\xi+r} |x(\xi) - x(\eta)|^2 d\eta + 2 \int_\xi^{\xi+r} |x(\eta)|^2 d\eta.$$

Consequently,

$$r|x(\xi)|^2 \leq 2r^2 \int_\xi^{\xi+r} \frac{|x(\xi) - x(\eta)|^2}{|\xi - \eta|^2} d\eta + 2 \int_\xi^{\xi+r} |x(\eta)|^2 d\eta,$$

which yields

$$|x(\xi)|^2 \leq 2r\|x\|_1 + \frac{2}{r} |x|_2^2.$$

The conclusion follows taking the minimum in $r$. $\qquad\square$

In §5.2 we will solve the stochastic equation (5.3) and in §5.3 and §5.4 we shall prove some useful estimates for the solution and its derivatives. §5 will be devoted to strong Feller property and irreducibility, §5.6 to existence of an invariant measure $\nu$ and §5.7 to solving the Kolmogorov equation in $L^2(H, \nu)$.

## 5.2 Solution of the stochastic differential equation

We follow here [38] and [50]. In order to define a concept of mild solution for equation (5.3), it is convenient to introduce for any $t > 0$ the linear mapping $K(t)$,

$$K(t) : H_\#^1 \to H, \quad x \mapsto K(t)x,$$

where

$$K(t)x = e^{tA} D_\xi x, \quad x \in H_\#^1.$$

**Lemma 5.4.** *For any $t > 0$ we have*

$$|K(t)x|_2 \leq \kappa t^{-\frac{3}{4}} |x|_1, \tag{5.7}$$

*for all $x \in H_\#^1$.*

By Lemma 5.4 it follows that for any $t > 0$ the mapping $K(t)$ can be uniquely extended to a bounded linear mapping (which we still denote by $K(t)$) from $L^1$ into $H$. Moreover, (5.7) holds for all $x \in L^1$.

*Proof of Lemma 5.4.* Fix $h \in H$ and $x \in H_\#^1$. Then we have, integrating by parts in $[0, 2\pi]$ and using the symmetry of $e^{tA}$,

$$\langle h, K(t)x \rangle = \langle e^{tA}h, D_\xi x \rangle = -\langle D_\xi(e^{tA}h), x \rangle,$$

since $e^{tA}x(0) = e^{tA}x(2\pi)$. Consequently, in view of (5.4),

$$|\langle h, K(t)x \rangle| \leq |D_\xi(e^{tA}h)|_\infty |x|_1 \leq \kappa t^{-\frac{3}{4}} |x|_1 |h|_2.$$

Now, (5.7) follows from the arbitrariness of $h$. $\qquad\square$

**Exercise 5.5.** Prove that the mapping $K(t)$ maps $L^1$ into $H_{\#}^{2\sigma}$ for all $\sigma \in (0, 1/4)$ and there exists $c_\sigma > 0$ such that

$$\|K(t)x\|_{2\sigma} \le c_\sigma t^{-\frac{3}{4}-\sigma} \, |x|_1.$$

By a *mild* solution of (5.3) in $[0, T]$ we mean a function $X(\cdot, x) \in C_W([0, T]; H)$ such that

$$X(t, x) = e^{tA}x + \frac{1}{2} \int_0^t K(t - s)(X^2(s, x)) ds + W_A(t), \quad t \in [0, T]. \qquad (5.8)$$

Now we prove, following [38], an existence and uniqueness result.

**Theorem 5.6.** *For any $x \in H$ and $T > 0$ there exists a unique mild solution $X \in C_W([0, T]; H)$ of equation (5.3). Moreover, setting $Y(t, x) = X(t, x) - W_A(t)$, the following estimate holds:*

$$|Y(t, x)|_2^2 \; + \; \int_0^t e^{8 \int_s^t |W_A(r)|_\infty^2 dr} \, \|Y(s, x)\|_1^2 \, ds \le e^{8 \int_0^t |W_A(s)|_\infty^2 ds} \, |x|_2^2$$

$$(5.9)$$

$$+2 \int_0^t e^{8 \int_s^t |W_A(r)|_\infty^2 dr} \, |W_A(s)|_\infty^4 \, ds.$$

*Proof.* The proof is divided into two steps.

*Step* 1. Local existence

Since the nonlinear equation (5.8) has a quadratic nonlinearity, it is not difficult to prove (by using the contraction principle on the Banach space $C_W([0, T]; H)$) the existence of a mild solution in a small random interval depending on $\omega \in \Omega$ (recall that $W_A(\cdot) \in C([0, T]; C([0, 2\pi]))$, $\mathbb{P}$–a.s.).

*Step* 2. Global existence.

Let $X(t, x)$ be the solution of (5.8) defined in a maximal interval $I$ and let $[0, t] \subset I$. It is enough to prove an a-priori estimate for $|X(t, x)|_2$. This will imply, by a standard argument, that $I = [0, T]$.

Setting $Y(t) = Y(t, x) = X(t, x) - W_A(t)$ equation (5.8) becomes

$$Y(t) = e^{tA}x + \int_0^t K(t - s)[(Y(s) + W_A(s))^2] \, ds,$$

which can be considered as the mild form of the deterministic evolution equation

$$\begin{cases} Y'(t) = AY(t) + \frac{1}{2} D_\xi[(Y(t) + W_A(t))^2], & t \ge 0, \\[2mm] Y(0) = x. \end{cases} \qquad (5.10)$$

We proceed now formally assuming that $Y(t)$ is a strict solution of (5.10) (If not we make a suitable approximation by more regular solutions).

Multiplying both sides of the first equation in (5.10) by $Y(t)$ and integrating with respect to $\xi$ over $[0, 2\pi]$ we get

$$\frac{1}{2}\frac{d}{dt}\int_0^{2\pi} Y^2(t)d\xi = \int_0^{2\pi} Y_{\xi\xi}(t)Y(t)d\xi$$

$$+\frac{1}{2}\int_0^{2\pi} D_\xi[(Y(t)+W_A(t))^2]Y(t)d\xi.$$

It follows, integrating by parts, that

$$\frac{1}{2}\frac{d}{dt}\int_0^{2\pi} Y^2(t)d\xi + \int_0^{2\pi} Y_\xi^2(t)d\xi = -\frac{1}{2}\int_0^{2\pi}(Y(t)+W_A(t))^2 Y_\xi(t)d\xi$$

$$= -\frac{1}{2}\int_0^{2\pi} Y^2(t)Y_\xi(t)d\xi - \int_0^{2\pi} W_A(t)Y(t)Y_\xi(t)d\xi - \frac{1}{2}\int_0^{2\pi} W_A^2(t)Y_\xi(t)d\xi.$$

Notice that

$$\int_0^{2\pi} Y^2(t)Y_\xi(t)d\xi = 0,$$

since $Y(t)$ is $2\pi$-periodic. So, the first term in the last identity disappears. Consequently, we have

$$\frac{1}{2}\frac{d}{dt}|Y(t)|_2^2 + \|Y(t)\|_1^2 \le |W_A(t)|_\infty\|Y(t)\|_1\,|Y(t)|_2 + \frac{1}{2}|W_A(t)|_\infty^2\|Y(t)\|_1$$

$$\le \frac{1}{2}\|Y(t)\|_1^2 + |W_A(t)|_\infty^4 + 4|W_A(t)|_\infty^2|Y(t)|_2^2.$$

In conclusion we find that

$$\frac{d}{dt}|Y(t)|_2^2 + \|Y(t)\|_1^2 \le 8|W_A(t)|_\infty^2|Y(t)|_2^2 + 2|W_A(t)|_\infty^4,$$

and thus, by a standard comparison result, (5.9) follows.

This is the required a priori bound. Now $Y$ can be continued up to $T$ to a function of $C_W([0,T];H)$, which implies that $X \in C_W([0,T];H)$ as claimed. $\square$

To make rigorous several computations in what follows, it is convenient to consider an approximating equation of (5.3). For any $m \in \mathbb{N}$ we define

$$b_m(x) = \frac{1}{2}P_m D_\xi\left(\frac{(P_m x)^2}{1+m^{-1}(P_m x)^2}\right), \qquad x \in H,$$

where

$$P_m = \sum_{i=-m}^{m} e_i \otimes e_i, \qquad m \in \mathbb{N}.$$

Then we consider the approximating problem

$$\begin{cases} dX_m(t) & = & (AX_m(t) + b_m(X_m(t))dt + \sqrt{C}\, dW(t), \\[2mm] X_m(0) & = & x. \end{cases} \tag{5.11}$$

The corresponding mild form is

$$X_m(t) = e^{tA}x + \frac{1}{2}\int_0^t K_m(t-s)\left(\frac{(P_m X(s))^2}{1+m^{-1}(P_m X(s))^2}\right)ds + W_A(t), \quad t \in [0,T],$$

where

$$K_m(t)x = P_m e^{tA} D_\xi x, \quad x \in H^1_\#.$$

Since, as easily checked, all previous estimates are uniform on $m$, we have the following result.

**Theorem 5.7.** *For any $x \in H$ and $T > 0$ there exists a unique mild solution $X_m \in C_W([0,T]; H)$ of equation (5.11). Moreover $X_m \to X$ in $C_W([0,T]; H)$.*

## 5.3   Estimates for the solutions

We follow here [32].

**Lemma 5.8.** *Let $m \in \mathbb{N}$ and let $X_m(t,x)$ be the solution of (5.11). Then we have*

$$|X_m(t,x)|_2^2 + 2\int_0^t \|X_m(s,x)\|_1^2 ds \tag{5.12}$$

$$= |x|_2^2 + t\,\mathrm{Tr}\, C_m + 2\int_0^t \langle X_m(s,x), \sqrt{C_m}\, dW_s\rangle,$$

*and*

$$\mathbb{E}\left(|X_m(t,x)|_2^2 + 2\int_0^t \|X_m(s,x)\|_1^2 ds\right) = |x|_2^2 + t\,\mathrm{Tr}\,[C_m], \tag{5.13}$$

*where $C_m = P_m C$.*

*Proof.* Identity (5.12) follows from Itô's formula. Taking expectation we find (5.13). □

**Proposition 5.9.** *Let $X(t,x)$ be the mild solution of (5.3). Then we have*

$$|X(t,x)|_2^2 + 2\int_0^t \|X(s,x)\|_1^2 ds = |x|_2^2 + t\,\mathrm{Tr}\, C + 2\int_0^t \langle X(s,x), \sqrt{C}\, dW_s\rangle, \tag{5.14}$$

*and*

$$\mathbb{E}\left(|X(t,x)|_2^2 + 2\int_0^t \|X(s,x)\|_1^2 ds\right) = |x|_2^2 + t\,\mathrm{Tr}\,[C]. \tag{5.15}$$

*Proof.* Identities (5.14), (5.15) follow from (5.12) and (5.13) respectively letting $m \to \infty$. $\square$

Now we want to estimate $\mathbb{E}\left(e^{\varepsilon|X(t,x)|_2^2}\right)$ for $\varepsilon > 0$ sufficiently small.

**Proposition 5.10.** *If $\varepsilon\|C\| \leq 1$ we have*

$$\mathbb{E}\left(e^{\varepsilon|X(t,x)|_2^2}\right) \leq e^{\varepsilon|x|_2^2 + \varepsilon t \, \mathrm{Tr} \, C}, \quad t \geq 0, \, x \in H. \tag{5.16}$$

*Proof.* Let us apply Itô's formula to

$$\varphi(x) = e^{\varepsilon|x|_2^2}, \quad x \in H.$$

Since

$$D\varphi(x) = 2\varepsilon x \varphi(x), \quad D^2\varphi(x) = 2\varepsilon\varphi(x) + 4\varepsilon^2 x \otimes x\varphi(x),$$

we have, fixing $m \in \mathbb{N}$ and setting $X_m(t,x) = X_m$,

$$\begin{aligned}
d\varphi(X_m) &= 2\varepsilon\varphi(X_m)\langle AX_m + b_m(X_m), X_m\rangle dt + \varepsilon \, \mathrm{Tr} \, [C_m]\varphi(X_m)dt \\
&\quad + 2\varepsilon^2\varphi(X_m)|\sqrt{C_m} \, X_m|_2^2 dt + 2\varepsilon\varphi(X_m)\langle X_m, \sqrt{C_m} \, dW_t\rangle.
\end{aligned}$$

It follows, taking expectation and recalling that $\langle b_m(X_m), X_m\rangle = 0$, that

$$\begin{aligned}
\mathbb{E}[\varphi(X_m(t))] &= \varphi(x) + \varepsilon \, \mathrm{Tr} \, C_m \int_0^t \mathbb{E}[\varphi(X_m(s))]ds \\
&\quad + 2\varepsilon\mathbb{E}\int_0^t \varphi(X_m(s))\left[\varepsilon|\sqrt{C_m} \, X_m(s)|_2^2 - \|X_m(s)\|_1^2\right]ds.
\end{aligned} \tag{5.17}$$

Since, by the Poincaré inequality,

$$\varepsilon|\sqrt{C_m} \, X_m(s)|_2^2 - \|X_m(s)\|_1^2 \leq \|C_m\| \, |X_m(s)|_2^2 - |X_m(s)|_2^2 \leq 0,$$

provided $\varepsilon\|C\| \leq 1$, we deduce by (5.17) that

$$\mathbb{E}[\varphi(X_m(t))] \leq \varphi(x) + \varepsilon \, \mathrm{Tr} \, C_m \int_0^t \mathbb{E}[\varphi(X_m(s))]ds.$$

Now from Gronwall's lemma it follows that

$$\mathbb{E}[\varphi(X_m(t))] \leq e^{\varepsilon t \mathrm{Tr} \, C_m},$$

and the conclusion follows letting $m \to \infty$. $\square$

We shall need an improvement of estimate (5.16).

**Proposition 5.11.** *Let* $\varepsilon \leq \varepsilon_0 = \frac{1}{\|C\|}$ *, then for any* $t \geq 0$ *and* $m \in \mathbb{N}$, *we have*

$$\mathbb{E}\left(e^{\varepsilon|X_m(t,x)|_2^2+\varepsilon\int_0^t \|X_m(s,x)\|_1^2 ds}\right) \leq e^{\varepsilon|x|^2+\varepsilon t \ \mathrm{Tr}\ C}, \quad x \in H,\ t \geq 0. \tag{5.18}$$

*Proof.* Fix $m \in \mathbb{N}$, set $X_m(t,x) = X_m(t)$ and

$$Z(t) = |X_m(t)|_2^2 + \int_0^t \|X_m(s)\|_1^2 ds,\ t \geq 0.$$

We have

$$\begin{aligned}
dZ(t) &= 2\langle X_m(t), AX_m(t) + b_m(X_m(t))\rangle dt + \mathrm{Tr}\ C_m dt \\
&\quad + 2\langle X_m(t), \sqrt{C_m}\ dW_t\rangle + \|X_m(t)\|_1^2 dt,
\end{aligned}$$

which implies

$$dZ(t) = \left(-2\|X_m(t)\|_1^2 + \mathrm{Tr}\ C_m\right) dt + 2\langle X_m, \sqrt{C_m}\ dW_t\rangle.$$

By Itô's formula applied to $e^{\varepsilon Z(t)}$, it follows that

$$\begin{aligned}
de^{\varepsilon Z(t)} &= \varepsilon e^{\varepsilon Z(t)} \left[-2\|X_m(t)\|_1^2 + \mathrm{Tr}\ C_m + 2\varepsilon^2|\sqrt{C_m}\ X_m|_2^2\right] dt \\
&\quad + 2\varepsilon e^{\varepsilon Z(t)}\langle X_m(t), \sqrt{C_m}\ dW_t\rangle.
\end{aligned}$$

Then, integrating in $t$ and taking expectation we obtain

$$\begin{aligned}
\mathbb{E}\left(e^{\varepsilon Z(t)}\right) &= e^{\varepsilon|x|_2^2} + \varepsilon\ \mathrm{Tr}\ C_m \int_0^t \mathbb{E}\left(e^{\varepsilon Z(s)}\right) ds \\
&\quad + 2\varepsilon \mathbb{E}\left(\int_0^t e^{\varepsilon Z(s)}(-2\|X_m(s)\|_1^2 + \varepsilon|\sqrt{C_m}\ X_m(s)|_1^2) ds\right).
\end{aligned} \tag{5.19}$$

Since as before $-2\|X_m(s)\|_1^2 + \varepsilon|\sqrt{C_m}\ X_m(s)|_1^2 \leq 0$, we have

$$\mathbb{E}\left(e^{\varepsilon Z(t)}\right) \leq e^{\varepsilon|x|_2^2} + \varepsilon\ \mathrm{Tr}\ C_m \int_0^t \mathbb{E}\left(e^{\varepsilon Z(s)}\right) ds.$$

Now, from Gronwall's lemma we obtain

$$\mathbb{E}\left(e^{\varepsilon|X_m(t,x)|_2^2+\varepsilon\int_0^t \|X_m(s,x)\|_1^2 ds}\right) \leq e^{\varepsilon|x|_2^2+\varepsilon t \ \mathrm{Tr}\ C}, \quad x \in H,\ t \geq 0,$$

and the conclusion follows letting $n \to \infty$. $\qquad\qquad\square$

## 5.4 Estimates for the derivative of the solution with respect to the initial datum

Let $h \in H$, setting $\eta_m^h(t, x) = DX_m(t, x)h$, then we have by Theorem 3.6,

$$\begin{cases} \dfrac{d}{dt}\,\eta_m^h(t, x) = A\eta_m^h(t, x) + D_\xi[X_m(t, x)\eta_m^h(t, x)], \\[2mm] \eta_m^h(0, x) = h. \end{cases} \tag{5.20}$$

**Lemma 5.12.** *There exists $c_1 > 0$ such that*

$$|\eta_m^h(t, x)|_2^2 + \int_0^t \|\eta_m^h(s, x)\|_1^2 ds \le e^{c_1 \int_0^t \|X_m(s,x)\|_1^{4/3} ds}\, |h|_2^2. \tag{5.21}$$

*Proof.* Multiplying scalarly both sides of (5.20) by $\eta_m^h(t, x)$, and integrating by parts with respect to $\xi$, yields

$$\frac{1}{2}\frac{d}{dt}\,|\eta_m^h(t, x)|_2^2 + \|\eta_m^h(t, x)\|_1^2$$

$$= \int_0^{2\pi} D_\xi(X_m(t, x)\eta_m^h(t, x))\eta_m^h(t, x)d\xi$$

$$= -\int_0^{2\pi} X_m(t, x)\eta_m^h(t, x)D_\xi\eta_m^h(t, x)d\xi$$

$$= \frac{1}{2}\int_0^{2\pi} (\eta_m^h(t, x))^2 D_\xi X_m(t, x)d\xi.$$

Now, using the interpolatory estimate (5.5), and recalling the Sobolev embedding $H_\#^{1/4} \subset L^4$, yields for a suitable constant $c > 0$,

$$\frac{1}{2}\frac{d}{dt}\,|\eta_m^h(t, x)|_2^2 + \|\eta_m^h(t, x)\|_1^2$$

$$\le \frac{1}{2}\,\|X_m(t, x)\|_1\,|\eta_m^h(t, x)|_4^2$$

$$\le \frac{c}{2}\,\|X_m(t, x)\|_1\,\|\eta_m^h(t, x)\|_{\frac{1}{4}}^2$$

$$\le \frac{c}{2}\,\|X_m(t, x)\|_1\,|\eta_m^h(t, x)|_2^{3/2}\,\|\eta_m^h(t, x)\|_1^{1/2}.$$

Using the elementary inequality

$$ab \le \frac{3}{4}\,a^{4/3} + \frac{1}{4}\,b^4, \quad a, b > 0,$$

with

$$a = \frac{c}{2} \, \|X(t,x)\| \, |\eta(t,x)|_2^{3/2}, \quad b = \|\eta(t,x)\|_1^{1/2},$$

yields

$$\frac{1}{2} \frac{d}{dt} \, |\eta_m^h(t,x))|_2^2 + \|\eta_m^h(t,x))\|_1^2$$

$$\leq \frac{3}{4} \left(\frac{c}{2}\right)^{4/3} \|X_m(t,x)\|_2^{4/3} |\eta_m^h(t,x)|_2^2 + \frac{1}{4} \|\eta_m^h(t,x))\|_1^2.$$

It follows that

$$\frac{d}{dt} \, |\eta_m^h(t,x)|_2^2 + \|\eta_m^h(t,x)\|_1^2 \leq c_1 \|X_m(t,x)\|_2^{4/3} |\eta(t,x)|_2^2,$$

which implies, by a classical comparison result,

$$|\eta_m^h(t,x)|_2^2 \leq e^{c_1 \int_0^t \|X_m(s,x)\|_1^{4/3} ds} \, |h|_2^2$$

$$- \int_0^t e^{c_1 \int_s^t \|X_m(\sigma,x)\|_2^{4/3} d\sigma} \|\eta_m^h(s,x))\|_1^2 ds \, |h|_2^2,$$

which yields (5.21).                                                                   $\square$

**Proposition 5.13.** *For any $\varepsilon \leq \varepsilon_0 = \frac{1}{\|C\|}$ there exists $c_{1,\varepsilon} > 0$ such that*

$$\mathbb{E} \left( |\eta^h(t,x)|_2^2 + \int_0^t \|\eta^h(s,x)\|_1^2 ds \right) \leq e^{\varepsilon |x|_2^2 + c_{1,\varepsilon} t} |h|_2^2, \quad t \geq 0, \, x, h \in H, \quad (5.22)$$

*and, for any $\sigma \in [0,1]$,*

$$\int_0^t \|\eta^h(s,x)\|_\sigma^2 \, ds \leq t^{1-\sigma} e^{\varepsilon |x|_2^2 + c_{1,\varepsilon} t} \, |h|_2^2, \quad t \geq 0, \, x, h \in H. \quad (5.23)$$

*Proof.* By Lemma 5.12 we have, letting $m \to \infty$,

$$|\eta^h(t,x)|_2^2 + \int_0^t \|\eta^h(s,x)\|_1^2 ds \leq e^{c_1 \int_0^t \|X(s,x)\|_1^{4/3} ds} \, |h|_2^2.$$

Clearly, for any $\varepsilon > 0$ there exists $c_\varepsilon > 0$ such that

$$c_1 \int_0^t \|X(s,x)\|_1^{4/3} ds \leq \varepsilon \int_0^t \|X(s,x)\|_1^2 ds + c_\varepsilon t,$$

so that

$$|\eta(t,x)|_2^2 + \int_0^t \|\eta(s,x)\|_1^2 \leq e^{\varepsilon \int_0^t \|X(s,x)\|_1^2 ds + c_\varepsilon t} \, |h|_2^2.$$

Now, by (5.18) it follows, taking expectation, that

$$\mathbb{E} \left( |\eta(t,x)|_2^2 + \int_0^t \|\eta(s,x)\|_1^2 ds \right) \leq e^{\varepsilon |x|_2^2 + \varepsilon t \, \mathrm{Tr} \, C + c_\varepsilon t} \, |h|_2^2,$$

and (5.22) follows. Finally, (5.23) follows by interpolation.                           $\square$

## 5.5 Strong Feller property and irreducibility

We start from the Bismut–Elworthy formula for the approximating semigroup $P_t^m$ (see Theorem 3.11),

$$\langle DP_t^m \varphi(x), h \rangle = \frac{1}{t} \, \mathbb{E} \left[ \varphi(X_m(t,x)) \int_0^t \langle (-A)^{\gamma/2} \eta_m^h(s,x), dW(s) \rangle \right],$$

where $x, h \in H$, $t > 0$. Using the Hölder inequality we find that

$$|\langle DP_t^m \varphi(x), h \rangle|^2 \leq t^{-2} \, \|\varphi\|_0^2 \, \mathbb{E} \left[ \int_0^t \|\eta_m^h(s,x)\|_\gamma^2 ds \right], \quad x, h \in H. \tag{5.24}$$

**Proposition 5.14.** *Assume that $\gamma \in [0,1)$. Then $P_t$ is strong Feller.*

*Proof.* By (5.24) and (5.23) it follows that for any $\varphi \in C_b(H)$,

$$|\langle DP_t^m \varphi(x), h \rangle|^2 \leq t^{-1-\sigma} \, e^{\epsilon |x|^2 + c_{1,\epsilon} t} \, \|\varphi\|_0 \, |h|_2^2, \quad x, h \in H.$$

Consequently, by the arbitrariness of $h$,

$$|DP_t^m \varphi(x)| \leq t^{-\frac{1+\sigma}{2}} e^{\frac{1}{2}(\epsilon |x|^2 + c_{1,\epsilon} t)} \, \|\varphi\|_0, \quad x \in H.$$

Letting $m \to \infty$ we obtain

$$|DP_t \varphi(x)| \leq t^{-\frac{1+\sigma}{2}} e^{\frac{1}{2}(\epsilon |x|^2 + c_{1,\epsilon} t)} \, \|\varphi\|_0, \quad x \in H.$$

Therefore the strong Feller property follows, arguing as in the proof of Step 2 of Theorem 3.11. $\qquad \square$

We prove now irreducibility of $P_t$. For this we consider the controlled system

$$\begin{cases} y'(t) = Ay(t) + b(y(t)) + (-A)^{-\gamma/2} u(t), & t \in [0.T], \\ y(0) = x \in H, \end{cases} \tag{5.25}$$

where $u \in L^2(0,T;H)$.

By a fixed point argument (as in the proof of Theorem 5.6) for any $x \in H$ and any $u \in L^2(0,T;H)$ there exists a unique mild solution $y = y(t,x;u)$ of (5.25), that is

$$y(t) = e^{tA} x + \frac{1}{2} \int_0^t K(t-s)(y(s)^2) ds + \int_0^t (-A)^{-\gamma/2} e^{(t-s)A} u(s) ds, \quad t \geq 0.$$

Moreover, if $x \in C_\#$ we have $y \in C([0,T]; C_\#)$.

We start as before with a result of approximate controllability.

**Proposition 5.15.** *Let $T > 0$, $x_0, x_1 \in H$, $\varepsilon > 0$. Then there exists $u \in C([0,T]; C_\#)$ such that*

$$|y(T, x_0; u) - x_1| \le \varepsilon. \tag{5.26}$$

*Proof.* For any $z_0, z_1 \in C_\#^2$ we set as before

$$\alpha_{z_0,z_1}(t) = \frac{T-t}{T} z_0 + \frac{t}{T} z_1, \quad t \in [0,T],$$

and

$$\beta_{z_0,z_1}(t) = \frac{d}{dt} \alpha_{z_0,z_1}(t) - A\alpha_{z_0,z_1}(t) - b(\alpha_{z_0,z_1}(t)), \quad t \in [0,T].$$

Note that $\beta_{z_0,z_1}(t) \in C_\#$ for all $t \in [0,T]$ and we have

$$\alpha_{z_0,z_1}(0) = z_0, \quad \alpha_{z_0,z_1}(T) = z_1.$$

Let now $x_0, x_1 \in H$ and $\varepsilon > 0$ be fixed. Choose $z_0, z_1 \in C_\#^2$ and $u \in C([0,T], C_\#)$ such that

(i) $|x_0 - z_0|_2 < c\varepsilon$, $\quad |x_1 - z_1|_2 < c\varepsilon$,

(ii) $|\beta_{z_0,z_1}(t) - (-A)^{-\gamma/2} u(t)|_2 \le c\varepsilon$, $\quad t \in [0,T]$,

where $c > 0$. We are going to show that the constant $c$ may be choosen such as $u$ fulfills (5.26). We have in fact

$$|y(T, x_0; u) - x_1|_2 \le |y(T, x_0; u) - y(T, z_0; u)|_2 + |y(T, z_0; u) - x_1|_2$$

$$\le |y(T, x_0; u) - y(T, z_0; u)|_2 + |y(T, z_0; u) - \alpha_{z_0,z_1}(T)|_w = I_1 + I_2. \tag{5.27}$$

Let us estimate $I_1$. Set $v(t) = y(t, x_0; u) - y(t, z_0; u)$, $t \in [0,T]$, then $v(t)$ is a solution of the integral equation

$$v(t) = e^{tA}(x_0 - z_0) + \frac{1}{2} \int_0^t K(t-s)[(y(t, x_0; u) - y(t, z_0; u))^2] ds.$$

Choose now $\delta$ such that

$$\sup_{t \in [0,T]} |y(t, x_0; u)|_2 + \sup_{t \in [0,T]} |y(t, z_0; u)|_2 \le \delta.$$

Then recalling Lemma 5.2, we have

$$|v(t)|_2 \le |x_0 - z_0|_2 + \frac{\kappa\delta}{2} \int_0^t (t-s)^{-\frac{3}{4}} |v(s)|_2 ds.$$

Thus, using a generalized Gronwall lemma we see that there exists $c_1 > 0$ such that

$$|I_1|_2 = |v(T)|_2 \le c_1 |x_0 - z_0|_2 \le cc_1 e^{2c_1 T} \varepsilon. \tag{5.28}$$

Let us estimate $I_2$. Set $w(t) = y(t, z_0; u) - \alpha_{z_0, z_1}(t)$, $t \in [0, T]$. Then $w$ is a solution of the integral equation

$$w(t) = \frac{1}{2} \int_0^t K(t-s)[(y(t, z_0; u) - \alpha_{z_0, z_1}(t))^2]ds.$$

Arguing as before we see that there exists $c_2 > 0$ such that

$$|I_2|_2 = |w(T)|_2 \leq T e^{(2c_1+1)T} \varepsilon^2 \tag{5.29}$$

and (5.26) follows from (5.27)–(5.29) . $\qquad\square$

We can now prove irreducibility following [50].

**Theorem 5.16.** *The transition semigroup $P_t$ is irreducible.*

*Proof.* To prove irreducibility it is enough to show that for all $x_0, x_1 \in H$ and $\varepsilon > 0$ we have

$$\mathbb{P}\left(|X(t, x_0) - x_1| \geq \varepsilon\right) < 1. \tag{5.30}$$

By Proposition 5.15 there is a control $u \in C([0, T]; C_\#)$ such that

$$|y(T, x_0; u) - x_1|_2 < \frac{\varepsilon}{2}. \tag{5.31}$$

We now proceed as before by comparing $X(t, x_0)$ and $y(t) = y(t, x_0; u)$. We have

$$|X(t, x) - y(t)|_2 \leq \frac{1}{2} \int_0^t |K(t-s)(X^2(s, x) - y^2(s))|_2 ds + |W_A(t) - \sigma_u(t)|_2, \tag{5.32}$$

where

$$\sigma_u(t) = \int_0^t (-A)^{-\gamma/2} e^{(t-s)A} u(s) ds, \quad t \in [0, T].$$

Recalling estimate (5.9) we see that there exists a positive continuous increasing function $\delta$ such that

$$|X(t, x)|_2 + |y(t)|_2 \leq \delta(\|W_A - \sigma_u\|_C),$$

where $C = C([0, T]; C_\#)$. Consequently, in view of Lemma 5.2, it follows that

$$|X(t, x) - y(t)|_2 \leq \frac{\kappa}{2} \delta(\|W_A - \sigma_u\|_C) \int_0^t |(t-s)^{-\frac{3}{4}}|X(s, x) - y(s))|_2 ds$$

$$+ |W_A(t) - \sigma_u(t)|_2.$$

Using the Gronwall lemma we see that there exists a positive constant $c$ such that

$$|X(t, x) - y(t)|_2 \leq c\delta(\|W_A - \sigma_u\|_C)\|W_A - \sigma_u\|_C. \tag{5.33}$$

Now we can conclude the proof.

We have in fact by (5.33), recalling (5.30), that

$$\mathbb{P}(|X(t,x_0) - x_1| \geq \varepsilon) \leq \mathbb{P}(|X(t,x_0) - y(T)| \geq \frac{\varepsilon}{2})$$

$$\leq \mathbb{P}\left(\delta(\|W_A - \sigma_u\|_C)\|W_A - \sigma_u\|_C > \frac{\varepsilon}{2c}\right) < 1,$$

since $W_A$ is full in $C([0,T]; C_\#)$, see Exercise 2.16.                    □

## 5.6   Invariant measure $\nu$

We shall denote by $P_t$ and $P_t^m$ the transition semigroups

$$P_t\varphi(x) = \mathbb{E}[\varphi(X(t,x))], \quad \varphi \in C_b(H),$$

and

$$P_t^m\varphi(x) = \mathbb{E}[\varphi(X_m(t,x))], \quad \varphi \in C_b(H),$$

where $X(t,x)$ and $X_m(t,x)$ are the solutions of (5.3) and (5.11) respectively.

By Theorem 5.7 it follows that for any $\varphi \in C_b(H)$ we have

$$\lim_{m \to \infty} P_t^m\varphi(x) = P_t\varphi(x), \quad t \geq 0, \ x \in H. \tag{5.34}$$

**Proposition 5.17.** $P_t^m$ *has a unique invariant measure* $\nu_n$. *Moreover,* $P_t$ *has an invariant measure* $\nu$ *which is a weak limit point of* $\{\nu_n\}$ *and it is full.*

*Proof.* Existence and uniqueness of the invariant measure $\nu_n$ of $P_t^m$ follow from finite dimensional classical results, see [71], since Ker $C_m = \{0\}$.

Let us prove existence of an invariant measure for $P_t$. Integrating both sides of (5.13) with respect to $\nu_m$ yields

$$\int_H \|x\|_1^2 \nu_m(dx) = \frac{1}{2} \text{Tr}\,[C_m] \leq \frac{1}{2} \text{Tr}\,C.$$

Now, let $R > 0$, $B_R = \{x \in H_\#^1 : \|x\|_1 \leq R\}$ and notice that $B_R$ is a compact subset of $H$, since the embedding $H_\#^1 \subset H$ is compact. By the Chebyshev inequality we have

$$\nu_n(B_R^c) \leq \frac{1}{R^2} \int_H \|x\|_1^2 \nu_m(dx) \leq \frac{1}{2R^2} \text{Tr}\,C.$$

Therefore $\{\nu_n\}$ is tight. Let $\nu$ be a weak limit point of $\{\nu_n\}$, then it is easy to see that $\nu$ is invariant for $P_t$.

Finally, $\nu$ is full since $P_t$ is irreducible.                    □

**Corollary 5.18.** *If* $\gamma \in (\frac{1}{2}, 1)$ *there exists a unique invariant measure* $\nu$ *for* $P_t$. *Moreover,* $\nu$ *is full.*

*Proof.* Since $P_t$ is strong Feller (Proposition 5.14) and irreducible (Theorem 5.30), the uniqueness follows from Theorem 1.12.                    □

### 5.6.1 Estimate of some integral with respect to $\nu$

Let us consider the Kolmogorov operator

$$K\varphi = \frac{1}{2} \text{Tr} \left[ CD^2\varphi \right] + \langle Ax + b(x), D\varphi \rangle, \quad \varphi \in \mathscr{E}_A(H).$$

Setting $\varphi(x) = e^{\varepsilon|x|_2^2}$, we find formally

$$K\varphi = \varepsilon \text{Tr} \left[ C \right] e^{\varepsilon^2|x|_2^2} + 2\varepsilon |C^{1/2}x|_2^2 \, e^{\varepsilon|x|_2^2} - 2\varepsilon \|x\|_1^2 \, e^{\varepsilon|x|_2^2}.$$

Now, if $\nu$ is invariant for $P_t$ we can show, approximating $\varphi$ by regular functions, that $\int_H N\varphi d\nu = 0$, so that

$$\int_H \|x\|_1^2 \, e^{\varepsilon|x|_2^2} d\nu = \frac{1}{2} \text{Tr} \left[ C \right] \int_H e^{\varepsilon|x|_2^2} d\nu + \varepsilon \int_H |C^{1/2}x|_2^2 \, e^{\varepsilon|x|_2^2} d\nu$$

$$\leq \frac{1}{2} \text{Tr} \left[ C \right] \int_H e^{\varepsilon|x|_2^2} d\nu + \varepsilon \|C\| \int_H e^{\varepsilon|x|_2^2} d\nu.$$

Therefore, if $\varepsilon < \varepsilon_0 := \frac{1}{\|C\|}$, we have

$$\int_H \|x\|_1^2 \, e^{\varepsilon|x|_2^2} d\nu \leq C_\varepsilon \int_H e^{\varepsilon|x|_2^2} d\nu, \tag{5.35}$$

for a suitable constant $C_\varepsilon$.

**Proposition 5.19.** *For any $\varepsilon < \varepsilon_0$ there exists $M_\varepsilon > 0$ such that*

$$\int_H e^{\varepsilon|x|_2^2} d\nu + \int_H \|x\|_1^2 \, e^{\varepsilon|x|_2^2} d\nu \leq M_\varepsilon.$$

*Proof.* By (5.35) we have in fact for any $R > 0$, using the Chebyshev inequality,

$$\int_H e^{\varepsilon|x|_2^2} d\nu = \int_{\|x\|_1 \leq R} e^{\varepsilon|x|_2^2} d\nu + \int_{\|x\|_1 > R} e^{\varepsilon|x|^2} d\nu$$

$$\leq e^{\varepsilon R^2} + \frac{1}{R^2} \int_H \|x\|_2^2 \, e^{\varepsilon|x|_2^2} d\nu \leq e^{\varepsilon R^2} + \frac{C_\varepsilon}{R^2} \int_H e^{\varepsilon|x|^2} d\nu$$

and the conclusion follows choosing $R$ sufficiently large. $\qquad\square$

We now estimate $\int_H \|x\|^4 d\nu(x)$ following again [32]. For this we need a lemma.

**Lemma 5.20.** *There exists $c > 0$ such that*

$$|\langle b(x), Ax \rangle| \leq c|x|_2^{10} + \frac{1}{2} |Ax|_2^2, \quad x \in D(A). \tag{5.36}$$

*Proof.* Let $x \in D(A)$. Then we have obviously

$$|b(x)|_2^2 = \int_0^{2\pi} |x(\xi)|^2 \, |D_\xi x(\xi)|^2 d\xi \leq |x|_\infty^2 \, \|x\|_1^2.$$

Recalling the Agmon estimate (5.6), we obtain

$$|b(x)|_2^2 \leq |x|_2 \, \|x\|_1^3.$$

Moreover, using the interpolatory estimate

$$\|x\|_1 \leq |x|_2^{\frac{1}{2}} \, |Ax|_2^{\frac{1}{2}}$$

yields

$$|b(x)|_2 \leq |x|_2^{\frac{5}{4}} \, |Ax|_2^{\frac{3}{4}}. \tag{5.37}$$

Consequently,

$$|\langle b(x), Ax \rangle| \leq |x|_2^{\frac{5}{4}} \, |Ax|_2^{\frac{7}{4}}.$$

Using the elementary estimate

$$ab \leq \frac{1}{7} \, a^8 + \frac{7}{8} \, b^{\frac{8}{7}}, \quad a, b \geq 0,$$

with $a = \epsilon |x|_2^{\frac{5}{4}}$ and $b = \epsilon^{-1} |Ax|_2^{\frac{7}{4}}$, we find

$$|\langle b(x), Ax \rangle| \leq \frac{1}{7} \, \epsilon^8 |x|_2^{10} + \frac{7}{8} \, \epsilon^{-\frac{8}{7}} |Ax|_2^2.$$

It is now enough to choose $\epsilon$ such that $\frac{7}{8} \, \epsilon^{-\frac{8}{7}} = \frac{1}{2}$. ☐

**Corollary 5.21.** *There exists $c_1 > 0$ such that*

$$|\langle b(x+z), Ax \rangle| \leq c(|x|_2^{10} + |z|_2^{10}) + \frac{1}{2} \, (|Ax|_2^2 + |Az|_2^2), \quad x \in D(A). \tag{5.38}$$

*Proof.* By (5.37) it follows that

$$|b(x+z)|_2 \leq |x+z|_2^{\frac{5}{4}} \, |Ax + Az|_2^{\frac{3}{4}}.$$

Consequently,

$$|\langle b(x+z), Ax \rangle| \leq |x+z|_2^{\frac{5}{4}} \, |Ax + Az|_2^{\frac{3}{4}} \, |Ax|_2 \leq k(|x|_2 + |z|_2)^{\frac{5}{4}} (|Ax|_2 + |Az|_2)^{\frac{7}{4}},$$

The conclusion follows arguing as before. ☐

Now we are in position to prove the following result.

**Proposition 5.22.** *There exists $c_1 > 0$ such that*

$$\int_H \|x\|_1^4 d\nu_n(x) \le c_1, \ n \in \mathbb{N}. \tag{5.39}$$

*Proof.* We start from the equation

$$\frac{d}{dt} Y(t) = AY(t) + b(Y(t) + W_A(t)). \tag{5.40}$$

Taking the inner product with $AY(t)$ and then multiplying by $\|Y(t)\|_1^2$, we find

$$\frac{1}{4} \frac{d}{dt} \|Y(t)\|_1^4 + \|Y(t)\|_1^2 |AY(t)|_2^2 \le |b(Y(t) + W_A(t))|_2 |AY(t)|_2 \|Y(t)\|_1^2.$$

By Corollary 5.21 we obtain

$$\frac{1}{4} \frac{d}{dt} \|Y(t)\|_1^4 + \|Y(t)\|_1^2 |AY(t)|_2^2$$

$$\le c \left(|Y(t)|_2^{10} + |W_A(t)|_2^{10}\right) \|Y(t)\|_1^2 + \frac{1}{2} \left(|AY(t)|_2^2 + |AW_A(t)|_2^2\right) \|Y(t)\|_1^2.$$

It follows that

$$\frac{1}{4} \frac{d}{dt} \|Y(t)\|_1^4 + \frac{1}{2} \|Y(t)\|_1^2 |AY(t)|_2^2$$

$$\le c(|Y(t)|_2^{10} + |W_A(t)|_2^{10})\|Y(t)\|_1^2 + \frac{1}{2} |AW_A(t)|_2^2 \|Y(t)\|_1^2.$$

Since $|Ax|_2 \ge \|x\|_1$ for all $x \in H$, we deduce that

$$\frac{d}{dt} \|Y(t)\|_1^4 + 2\|Y(t)\|_1^4$$

$$\le 4c(|Y(t)|_2^{10} + |W_A(t)|_2^{10})\|Y(t)\|_1^2 + 2 |AW_A(t)|_2^2\|Y(t)\|_1^2$$

$$\le \|Y(t)\|_1^4 + 8c^2(|Y(t)|_2^{10} + |W_A(t)|_2^{10})^2 + 2|AW_A(t)|_2^2.$$

Therefore

$$\frac{d}{dt} \|Y(t)\|_1^4 \le -\|Y(t)\|_1^4 + 8c^2(|Y(t)|_2^{10} + |W_A(t)|_2^{10})^2 + 2|AW_A(t)|_2^2.$$

By a classical comparison result, it follows that

$$\|Y(t)\|_1^4 \le e^{-t}\|x\|_1^4$$

$$+ \int_0^t e^{-(t-s)} \left(8c^2(|Y(s)|_2^{10} + |W_A(s)|_2^{10})^2 + 2|AW_A(s)|_2^2 \, ds\right)$$

which implies

$$\|X(t,x)\|_1^4 \leq 8e^{-t}\|x\|_1^4 + 8\|W_A(t)\|_1^4$$

$$+8\int_0^t e^{-(t-s)} \left(8c^2(2^9|X(s,x)|_2^{10} + 2^9|W_A(s))|_2^{10}\right.$$

$$+|W_A(s)|_2^{10})^2 + 2|AW_A(s)|_2^2 \Big) ds.$$

Taking expectation yields

$$P_t(\| \cdot \|_1^4)(x) \leq 8e^{-t}P_t(\| \cdot \|_1^4)(x) + 8 \, \mathbb{E}(\|W_A(t)\|_1^4)$$

$$+8\int_0^t e^{-(t-s)} \left(8c^2(2^9(P_s(| \cdot |_2^{10})(x) + 2^9\mathbb{E}(|W_A(s)|_2^{10}))^2 + 2\mathbb{E}(|AW_A(s)|_2^2) \, ds\right).$$

Now the conclusion follows by integrating with respect to $\nu$ and noting that

$$\mathbb{E}(\|W_A(t)\|_1^4) + \mathbb{E}(|W_A(t)|_2^{10}) + \mathbb{E}(|AW_A(t)|_2^2) \leq Ce^{-t},$$

by a suitable constant $C$, since $W_A(t)$ and $AW_A(t)$ are Gaussian random variables of covariance $(-A)^{-1-\gamma}$ and $(-A)^{-\gamma}$ respectively.                    □

**Corollary 5.23.** *There exists $\kappa > 0$ such that*

$$\int_H |b(x)|_2^2 \, \nu(dx) \leq \kappa. \tag{5.41}$$

*Proof.* We have in fact, for any $x \in H_\#^1$, recalling the Agmon estimate (5.6),

$$|b(x)|_2^2 \quad \leq \quad \frac{1}{4} \int_0^{2\pi} |x(\xi)|^2 \, |D_\xi x(\xi)|^2 d\xi \leq c|x|_\infty^2 \, \|x\|_1^2$$

$$\leq \quad 4c|x|_2 \, \|x\|_1^3 \leq \frac{c}{4} \, |x|_2^4 + \frac{3c}{4} \, \|x\|_1^4.$$

Therefore

$$\int_H |b(x)|_2^2 \, \nu(dx) \leq \frac{c}{4} \int_H |x|_2^4 \, \nu(dx) + \frac{3c}{4} \int_H \|x\|_1^4 \, \nu(dx) < +\infty,$$

by Proposition 5.19 and (5.16).                    □

## 5.7   Kolmogorov equation

We are here concerned with the semigroup $P_t$ in $L^2(H, \nu)$ where $\nu$ is an invariant measure. We know by Corollary 5.18 that if $\gamma \in (\frac{1}{2}, 1)$, $\nu$ is unique.

We denote as usual by $K_2$ the infinitesimal generator of $P_t$ in $L^2(H, \nu)$ and consider the Kolmogorov operator

$$K_0\varphi(x) = L\varphi + \langle b(x), D\varphi(x) \rangle, \quad \varphi \in \mathscr{E}_A(H),$$

where $L$ is the Ornstein–Uhlenbeck generator

$$L\varphi(x) = \frac{1}{2} \operatorname{Tr} [CD^2\varphi(x)] + \langle Ax, D\varphi(x) \rangle, \quad \varphi \in D(L, C_b(H)).$$

**Proposition 5.24.** *For any $\varphi \in \mathscr{E}_A(H)$ we have $\varphi \in D(K_2)$ and $K_2\varphi = K_0\varphi$.*

*Proof.* Let $\varphi \in \mathscr{E}_A(H)$, $h > 0$. By Itô's formula it follows that

$$P_h\varphi(x) - \varphi(x) = \mathbb{E} \int_0^h K_0\varphi(X(s, x))ds, \quad x \in H. \tag{5.42}$$

Consequently

$$\lim_{h \to 0} \frac{1}{h}(P_h\varphi(x) - \varphi(x)) = K_0\varphi(x), \quad x \in H,$$

pointwise. To prove that $\varphi \in D(K_2)$, it is enough to show that

$$\frac{1}{h}(P_h\varphi - \varphi), \quad h \in (0, 1],$$

is equibounded in $L^2(H, \nu)$.

It is clear that there exist $a, b > 0$ (depending on $\varphi$) such that

$$|L\varphi(x)| \le (a + b|x|_2), \quad x \in H.$$

Now, by (5.42) we have for any $x \in H$,

$$|P_h\varphi(x) - \varphi(x)| \le \int_0^h \mathbb{E}[a + b|X(s, x)|_2 + \|\varphi\|_1 \, |b(X(s, x))|_2]ds.$$

By the Hölder inequality we find

$$|P_h\varphi(x) - \varphi(x)|^2 \le h \int_0^h [\mathbb{E}(a + b|X(s, x)|_2 + \|\varphi\|_1 \, |b(X(s, x))|_2)]^2 ds$$

$$\le 2h \int_0^h \mathbb{E}[(a + b|X(s, x)|_2^2)ds + 2h\|\varphi\|_1^2 \int_0^h \mathbb{E}[|b(X(s, x))|_2^2]ds$$

$$= 2h \int_0^h P_s(a + b| \cdot |_2^2)(x)ds + 2\|\varphi\|_1^2 h \int_0^h P_s(|b(\cdot)|_2)(x)ds.$$

Integrating with respect to $\nu$ over $H$, and taking into account the invariance of $\nu$, yields

$$\|P_h\varphi - \varphi\|_{L^2(H, \nu)}^2 \le 2h^2 \int_H [(a + b|x|_2^2) + \|\varphi\|_1^2 \, |b(x)|_2^2]\nu(dx) < +\infty,$$

thanks to (5.41). Consequently, $\frac{1}{h}(P_h\varphi-\varphi)$ is equibounded in $L^2(H,\nu)$ as claimed.

□

To prove that $K_2$ is the closure of $K_0$ it is convenient to introduce the approximating problems

$$\begin{cases} dX_\varepsilon(t) = AX_\varepsilon(t)dt + b_\varepsilon(X_\varepsilon(t))dt + \sqrt{C}dW(t), \\ \\ X_\varepsilon(0) = x \in H, \end{cases} \tag{5.43}$$

where

$$b_\varepsilon(x) = \begin{cases} b(x) & \text{if } \|x\|_1 \leq \dfrac{1}{\varepsilon}, \\ \\ \dfrac{b(x)}{\varepsilon^2\|x\|_1^2} & \text{if } \|x\|_1 > \dfrac{1}{\varepsilon}. \end{cases} \tag{5.44}$$

Since, for $\varepsilon > 0$, $b_\varepsilon$ is regular and bounded, problem (5.44) has a unique solution which we denote by $X_\varepsilon(t,x)$. Notice that

$$|b_\varepsilon(x)|_2 \leq |b(x)|_2^2, \quad x \in H_\#^1.$$

Now we are ready to prove,

**Theorem 5.25.** $K_2$ is the closure of $K_0$.

*Proof.* We fix $\lambda > 0$, $f \in \mathscr{E}_A(H)$ and consider the approximating equation

$$\lambda\varphi_\varepsilon(x) - L\varphi_\varepsilon(x) - \langle b_\varepsilon(x), D\varphi_\varepsilon(x)\rangle = f(x). \tag{5.45}$$

Equation (5.45) has a unique solution $\varphi_\varepsilon \in C_b^1(H) \cap D(L)$. Moreover for any $h \in H$,

$$\langle D\varphi_\varepsilon(x), h\rangle = \int_0^{+\infty} e^{-\lambda t}\mathbb{E}[\langle Df(X_\varepsilon(t,x)), \eta_\varepsilon^h(t,x)\rangle dt.$$

Now, by Proposition 5.13 there exists $c > 0$ such that

$$|D\varphi_\varepsilon(x)| \leq \frac{1}{\lambda - c} e^{c|x|^2} \|f\|_1, \quad x \in H. \tag{5.46}$$

Let us show that $\varphi_\varepsilon \in D(\overline{K_0})$, where $\overline{K_0}$ is the closure of $K_0$. By Proposition 2.68 there exists a three-index sequence $\{\varphi_{\varepsilon,k_1,k_2,k_3}\} \subset \mathscr{E}_A(H)$ such that, for any $x \in H$,

$$\lim_{k_1\to\infty}\lim_{k_2\to\infty} \varphi_{\varepsilon,k_1,k_2,k_3}(x) = \varphi_\varepsilon(x),$$

$$\lim_{k_1\to\infty}\lim_{k_2\to\infty} L\varphi_{\varepsilon,k_1,k_2,k_3}(x) = L\varphi_\varepsilon(x),$$

$$\lim_{k_1\to\infty}\lim_{k_2\to\infty} D\varphi_{\varepsilon,k_1,k_2,k_3}(x) = D\varphi_\varepsilon(x),$$

and

$$|\varphi_{\varepsilon,k_1,k_2,k_3}(x)| + |L\varphi_{\varepsilon,k_1,k_2,k_3}(x)| + |D\varphi_{\varepsilon,k_1,k_2,k_3}(x)| \le a_\varepsilon(1+|x|), \quad k \in \mathbb{N}.$$

Then we have

$$K_0\varphi_{\varepsilon,k_1,k_2,k_3}(x) = L\varphi_{\varepsilon,k_1,k_2,k_3}(x) + \langle b(x), D\varphi_{\varepsilon,k_1,k_2,k_3}(x)\rangle.$$

We claim that

$$\lim_{k_1\to\infty}\lim_{k_2\to\infty}\lim_{k_3\to\infty} K_0\varphi_{\varepsilon,k_1,k_2,k_3}(x) = L\varphi_\varepsilon(x) + \langle b(x), D\varphi_\varepsilon(x)\rangle \quad \text{in } L^2(H,\nu).$$

The claim follows from the dominated convergence theorem since

$$|\langle b(x), D\varphi_\varepsilon(x)\rangle| \le a_\varepsilon(1+|x|)|b(x).$$

The claim implies that $\varphi_\varepsilon \in D(\overline{K_0})$ and that

$$\lambda\varphi_\varepsilon(x) - \overline{K_0}\varphi_\varepsilon(x) = f(x) + \langle b_\varepsilon(x) - b(x), D\varphi_\varepsilon(x)\rangle. \tag{5.47}$$

We prove finally that

$$\lim_{\varepsilon\to 0}\langle b_\varepsilon(x) - b(x), D\varphi(x)\rangle = 0 \quad \text{in } L^2(H,\nu). \tag{5.48}$$

This follows from the dominated convergence theorem since, in view of (5.46), Proposition 5.19 and Proposition 5.22,

$$\int_H |\langle b(x) - b_\varepsilon(x), D\varphi_\varepsilon(x)\rangle|^2\nu(dx) \le \frac{1}{\lambda-c}\|f\|_1\int_H |b(x)|_2^2 e^{c|x|_2^2}\nu(dx)$$

$$\le \frac{1}{\lambda-c}\|f\|_1\int_H |x|_2^2\,\|x\|_1^3 e^{c|x|^2}\,\nu(dx)$$

$$\le \frac{1}{\lambda-c}\|f\|_1\left(\int_H \frac{c}{4}\,|x|_2^4 e^{c|x|_1^2}\nu(dx) + \frac{3c}{4}\int_H \|x\|^4\nu(dx)\right).$$

Now, (5.48) implies the conclusion from the Lumer–Phillips Theorem 3.20. $\qquad\square$

# Chapter 6

# The Stochastic 2D Navier–Stokes Equation

## 6.1 Introduction and preliminaries

We are here concerned with the *Navier–Stokes* equation in $\mathcal{O} = [0, 2\pi] \times [0, 2\pi]$ perturbed by noise. For the sake of simplicity we shall only consider periodic boundary conditions and coulored noise. For Dirichlet boundary conditions see [8], for the $2D$ Navier–Stokes equation driven by white noise see [33].

We denote by $\xi = (\xi_1, \xi_2)$ the generic point of $\mathbb{R}^2$ and by

$$\langle \xi, \eta \rangle = \xi_1 \eta_1 + \xi_2 \eta_2, \quad \xi, \eta \in \mathbb{R}^2,$$

the scalar product in $\mathbb{R}^2$.

Let us consider the space $L_\#^2$ of all real $2\pi$-periodic functions in $\xi_1$ and $\xi_2$ which are measurable and square integrable on $\mathcal{O}$, endowed with the usual scalar product $\langle \cdot, \cdot \rangle_2$ and norm $|\cdot|_2$.

We also consider the space $(L_\#^2)^2$ consisting of all pairs $x = \begin{pmatrix} x^1 \\ x^2 \end{pmatrix}$ of real periodic functions in $\mathbb{R}^2$ endowed with the inner product [1]

$$\langle x, y \rangle_2 = \int_\mathcal{O} [x^1(\xi) y^1(\xi) + x^2(\xi) y^2(\xi)] d\xi, \quad x, y \in (L_\#^2)^2.$$

Moreover, for any $x \in (L_\#^2)^2$ we set $|x|_2 = (\langle x, x \rangle_2)^{\frac{1}{2}}$. It is convenient to consider also the complexified space $(L_\#^2)_{\mathbb{C}}^2$, of $(L_\#^2)^2$.

---

[1] Using the same symbol for inner product in $L_\#^2$ and in $(L_\#^2)^2$ will not cause any confusion.

We are concerned with the stochastic Navier–Stokes equation

$$
\begin{cases}
dZ = (\Delta_\xi Z - Z + D_\xi Z \cdot Z)dt + \nabla p dt + \sqrt{C}\, dV_t & \text{in } [0, +\infty) \times \mathcal{O}, \\[2mm]
\operatorname{div} Z = 0 & \text{in } [0, +\infty) \times \mathcal{O}, \\[2mm]
Z(t, \cdot) \text{ is periodic with period } 2\pi, \\[2mm]
Z(0, \cdot) = z & \text{in } \mathcal{O}.
\end{cases}
\tag{6.1}
$$

The unknown $Z = (Z_1, Z_2)$ represents the velocity and $p$ the pressure of a fluid, $C$ is a linear, symmetric, positive trace class operator on the Hilbert space $(L_\#^2)^2$ and $V_t$ is a cylindrical Wiener process in a probability space $(\Omega, \mathscr{F}, \mathbb{P})$ in $(L_\#^2)^2$.

Let us project, as usual, equation (6.1) on the space of square integrable divergence free vectors which we shall denote by $H$. Let us recall the definition of $H$. A complete orthonormal system on $(L_\#^2)_{\mathbb{C}}^2$ is given by

$$
f_{h,k} = \begin{pmatrix} e_h \\ e_k \end{pmatrix}, \quad h, k \in \mathbb{Z}^2
$$

where

$$
e_k(\xi) = \frac{1}{2\pi}\, e^{i\langle k, \xi\rangle}, \quad k = (k_1, k_2) \in \mathbb{Z}^2, \ \xi = (\xi_1, \xi_2),
$$

and $\langle k, \xi \rangle = k_1 \xi_1 + k_2 \xi_2$.

The closed subspace of $(L_\#^2)_{\mathbb{C}}^2$, of all *divergence free vectors* is defined by

$$
H_{\mathbb{C}} = \operatorname{span}\{g_k : k \in \mathbb{Z}^2\},
$$

where

$$
g_k = \frac{k^\perp}{|k|}\, e_k = \begin{pmatrix} -\frac{k_2}{|k|}\, e_k \\[2mm] \frac{k_1}{|k|}\, e_k \end{pmatrix}, \qquad k^\perp = \begin{pmatrix} -k_2 \\ k_1 \end{pmatrix}, \quad k \in \mathbb{Z}^2.
$$

Notice that

$$
\operatorname{div} g_k(\xi) = \frac{i}{|k|}\, e_k(-k_1 k_2 + k_1 k_2) = 0, \quad k \in \mathbb{Z}^2.
$$

Now we denote by $H$ the space of all real parts of element of $H_{\mathbb{C}}$ and by $\mathscr{P}$ the orthogonal projector of $(L_\#^2)^2$ onto $H$.

Then setting $X(t, x) = \mathscr{P}Z(t, x)$, $\mathscr{P}z = x$, $W(t) = \mathscr{P}V(t)$ and choosing $C$ as a linear operator in $H$, we obtain the following problem (where the pressure has disappeared):

$$
\begin{cases}
dX = (\mathscr{P}(\Delta X - X) + \mathscr{P}(D_\xi X \cdot X))dt + \sqrt{C}\, dW_t & \text{in } [0, +\infty) \times \mathcal{O}, \\[2mm]
X(t, \cdot) \text{ is periodic with period } 2\pi, \\[2mm]
X(0, \cdot) = x & \text{in } \mathcal{O}.
\end{cases}
\tag{6.2}
$$

### 6.1.1   The abstract setting

We first notice that any element $x = \begin{pmatrix} x^1 \\ x^2 \end{pmatrix}$ of $H_C$ can be developed in the Fourier series

$$x = \sum_{k \in \mathbb{Z}^2} x_k g_k, \tag{6.3}$$

where

$$x_k = \langle x, g_k \rangle_2 = -\frac{k_2}{|k|} \langle x^1, e_k \rangle_2 + \frac{k_1}{|k|} \langle x^2, e_k \rangle_2$$
$$= \frac{1}{2\pi|k|} \int_{\mathcal{O}} [-k_2 x^1(\eta) + k_1 x^2(\eta)] e^{-i\langle k, \eta \rangle} d\eta. \tag{6.4}$$

For any $\sigma \geq 0$ we define

$$\|x\|_\sigma^2 = \sum_{k \in \mathbb{Z}^2} (1 + |k|^2)^{\sigma/2} |x_k|^2$$

and set

$$H_\#^\sigma = \{x \in H : \|x\|_\sigma^2 < +\infty\}.$$

$H_\#^\sigma$, endowed with the norm $\|\cdot\|_\sigma$, is a Hilbert space. When $\sigma = 0$, $H_\#^\sigma$ reduces to $H$ and we set $\|x\|_0 = |x|_2$. Moreover, we shall denote by $L_\#^p$, $p \geq 1$, the subspace of $(L_\#^p)^2$ of all divergence free vectors and by $|\cdot|_p$ the norm in $L_\#^p$.

Let us define the *Stokes* operator $A \colon D(A) \to H$ setting

$$Ax = \mathscr{P}(\Delta_\xi x - x), \quad x \in D(A) = H_\#^2 \tag{6.5}$$

and denote by $A_C$ the complexification of $A$. $A_C$ is self-adjoint in $H$ and

$$Ag_k = -(1 + |k|^2)g_k, \quad k \in \mathbb{Z}^2. \tag{6.6}$$

We also define the nonlinear operator $b$ setting [2]

$$b(x) = \mathscr{P}(D_\xi x \cdot x), \quad x \in H_\#^1. \tag{6.7}$$

Now, problem (6.2) can be written in a compact form as

$$\begin{cases} dX &= (AX + b(X))dt + \sqrt{C}\, dW(t), \\ X(0) &= x. \end{cases} \tag{6.8}$$

We shall assume in the chapter that $\mathrm{Tr}\,[C(-A)] < \infty$. Moreover, for the sake of simplicity, we choose $C = (-A)^{-\gamma}$ with $\gamma > 2$ so that this assumption is fulfilled.

In §6.2 we will follow [50] and from §6.3 on [9].

---

[2] It is more commonly written $b(x) = \mathscr{P}(x \cdot \nabla x)$.

## 6.1.2   Basic properties of the nonlinear term

It is useful to introduce a trilinear form on $H_\#^1 \times H_\#^1 \times H_\#^1$ by setting

$$b(x,y,z) = \int_{\mathscr{O}} \langle D_\xi y \cdot x, z \rangle d\xi = \sum_{h,k=1}^{2} \int_{\mathscr{O}} D_k y_h \, x_k \, z_h \, d\xi, \quad x,y,z \in H_\#^1,$$

where $D_k$ represents the partial derivative with respect to $\xi_k$, $k = 1,2$. By the Hölder inequality it follows that

$$|b(x,y,z)| \leq 4|x|_4 \, \|y\|_1 \, |z|_4, \quad x,y,z \in H_\#^1. \tag{6.9}$$

By (6.9) it follows that $b$ maps : $H_\#^1$ into $L_\#^4$.

**Proposition 6.1.** *The following identities hold:*

$$b(x,y,z) = -b(x,z,y), \quad x,y,z \in H_\#^1, \tag{6.10}$$

$$b(x,y,y) = 0, \quad x,y \in H_\#^1. \tag{6.11}$$

*Proof.* We have in fact

$$b(x,y,z) \;=\; \sum_{h,k=1}^{2} \int_{\mathscr{O}} D_k y_h \, x_k \, z_h \, d\xi$$

$$=\; -\sum_{h,k=1}^{2} \int_{\mathscr{O}} y_h \, D_k x_k \, z_h \, d\xi - \sum_{h,k=1}^{2} \int_{\mathscr{O}} y_h \, x_k \, D_k z_h \, d\xi$$

$$=\; \int_{\mathscr{O}} \operatorname{div} x \, \langle y, z \rangle d\xi - b(x,z,y),$$

and (6.10) follows since div $x = 0$. Finally, (6.11) follows immediately from (6.10). $\qquad\square$

**Proposition 6.2.** *The operator $b$ can be uniquely extended to a continuous mapping from $L_\#^4$ into $H_\#^{-1}$, still denoted by $b$. Moreover,*

$$\|b(x) - b(y)\|_{-1} \leq 4(|x|_4 + |y|_4) \, |x - y|_4, \quad x,y \in L_\#^4. \tag{6.12}$$

*Proof.* Let $x,y,z \in H_\#^1$. Then, taking into account Proposition 6.1, we have

$$\langle b(x) - b(y), z \rangle \;=\; b(x,x,z) - b(y,y,z) = -b(x,z,x) + b(y,z,y)$$

$$=\; b(x,z,-x) + b(y,z,y) = b(x,z,y-x) + b(y-x,z,y).$$

Consequently, by (6.9) it follows that

$$\langle b(x) - b(y), z \rangle \leq 4[|x - y|_4 \, |x|_4 \, \|z\|_1 + |x - y|_4 \, |y|_4 \, \|z\|_1],$$

which yields the conclusion. $\qquad\square$

Next result, typical of periodic boundary conditions, is classical, see [3] and [60]. It will play an important rôle in what follows.

**Proposition 6.3.** *The following identity holds:*

$$b(x, x, Ax) = 0, \quad x \in D(A). \tag{6.13}$$

*Proof.* Let $x = \begin{pmatrix} x^1 \\ x^2 \end{pmatrix} \in H_{\mathbb{C}}$. Then by (6.3) and (6.4) we have

$$x^1 = -\sum_{k \in \mathbb{Z}^2} \frac{k_2}{|k|} x_k e_k, \quad x^2 = \sum_{k \in \mathbb{Z}^2} \frac{k_1}{|k|} x_k e_k.$$

Therefore

$$D_1 x^1 = -i \sum_{k \in \mathbb{Z}^2} \frac{k_1 k_2}{|k|} x_k e_k, \quad D_2 x^1 = -i \sum_{k \in \mathbb{Z}^2} \frac{k_2^2}{|k|} x_k e_k,$$

$$D_1 x^2 = i \sum_{k \in \mathbb{Z}^2} \frac{k_1^2}{|k|} x_k e_k, \quad D_2 x^2 = i \sum_{k \in \mathbb{Z}^2} \frac{k_1 k^2}{|k|} x_k e_k$$

and

$$D_\xi x = \begin{pmatrix} D_1 x^1 & D_2 x^1 \\ D_1 x^2 & D_2 x^2 \end{pmatrix} = \begin{pmatrix} -i \sum_{k \in \mathbb{Z}^2} \dfrac{k_1 k_2}{|k|} x_k e_k & -i \sum_{k \in \mathbb{Z}^2} \dfrac{k_2^2}{|k|} x_k e_k \\ i \sum_{k \in \mathbb{Z}^2} \dfrac{k_1^2}{|k|} x_k e_k & i \sum_{k \in \mathbb{Z}^2} \dfrac{k_1 k^2}{|k|} x_k e_k \end{pmatrix}.$$

Therefore $b(x) = D_\xi x \cdot x$ is given by

$$D_\xi x \cdot x = \begin{pmatrix} -i \sum_{h \in \mathbb{Z}^2} \dfrac{h_1 h_2}{|h|} x_h e_h & -i \sum_{h \in \mathbb{Z}^2} \dfrac{h_2^2}{|h|} x_h e_h \\ i \sum_{h \in \mathbb{Z}^2} \dfrac{h_1^2}{|h|} x_h e_h & i \sum_{h \in \mathbb{Z}^2} \dfrac{h_1 h_2}{|h|} x_h e_h \end{pmatrix} \begin{pmatrix} -\sum_{k \in \mathbb{Z}^2} \dfrac{k_2}{|k|} x_k e_k \\ \sum_{k \in \mathbb{Z}^2} \dfrac{k_1}{|k|} x_k e_k \end{pmatrix}$$

$$= \begin{pmatrix} i \sum_{h,k \in \mathbb{Z}^2} \dfrac{h_1 h_2 k_2 - h_2^2 k_1}{|h|\,|k|} x_h x_k e_{h+k} \\ i \sum_{h,k \in \mathbb{Z}^2} \dfrac{-h_1^2 k_2 + h_1 h_2 k_1}{|h|\,|k|} x_h x_k e_{h+k} \end{pmatrix}.$$

Consequently

$$\langle b(x), Ax \rangle = \begin{pmatrix} i \sum_{h,k \in \mathbb{Z}^2} \dfrac{h_1 h_2 k_2 - h_2^2 k_1}{|h|\,|k|} x_h x_k e_{h+k} \\ i \sum_{h,k \in \mathbb{Z}^2} \dfrac{-h_1^2 k_2 + h_1 h_2 k_1}{|h|\,|k|} x_h x_k e_{h+k} \end{pmatrix}$$

$$\times \begin{pmatrix} \sum_{m \in \mathbb{Z}^2} \dfrac{m_2}{|m|} (1+|m|^2) x_m e_m \\ - \sum_{m \in \mathbb{Z}^2} \dfrac{m_1}{|m|} (1+|m|^2) x_m e_m \end{pmatrix}$$

$$= i \sum_{\substack{h,k, \\ m \in \mathbb{Z}^2}} \frac{1+|m|^2}{|h|\,|k|\,|m|} (h_1 h_2 k_2 m_2 - h_2^2 k_1 m_2 + h_1^2 k_2 m_1 - h_1 h_2 k_1 m_1) x_h x_k \overline{x_m} e_{h+k-m}.$$

Now, integrating this identity in $\mathcal{O}$, and taking into account that the integrals of terms including $e_{h+k-m}$ vanish if $m \neq h + k$, yields

$$\langle b(x), Ax \rangle_2 = \frac{i}{2\pi} \sum_{h,k \in \mathbb{Z}^2} \frac{1+|h+k|^2}{|h|\,|k|\,|h+k|}$$

$$\times [h_1 h_2 k_2 (h_2+k_2) - h_2^2 k_1 (h_2+k_2) + h_1^2 k_2 (h_1+k_1) - h_1 h_2 k_1 (h_1+k_1)] x_h x_k \overline{x_{h+k}}.$$

Now, exchanging $k_1$ with $k_2$ and $h_1$ with $h_2$ yields

$$\langle b(x), Ax \rangle_2 = - \langle b(x), Ax \rangle_2$$

so that $\langle b(x), Ax \rangle_2 = 0$.                                                                 $\square$

### 6.1.3   Sobolev embedding and interpolatory estimates

We recall the Sobolev embedding theorem. For any $\epsilon > 0$ we have

$$H_\#^\epsilon \subset \begin{cases} L^{\frac{2}{1-\epsilon}} & \text{if } \epsilon \in (0,1), \\ C_\#(\mathbb{R}) & \text{if } \epsilon > 1/2, \end{cases}$$

with continuous inclusions. In particular, for $\epsilon = \frac{1}{2}$ we obtain the inclusion

$$H_\#^{\frac{1}{2}} \subset L_\#^4.$$

Therefore there exists $c_1 > 0$ such that

$$|x|_4 \leq c_1 \|x\|_{\frac{1}{2}}, \quad x \in H_\#^{\frac{1}{2}}. \tag{6.14}$$

We shall use the classical interpolatory estimate,

$$\|x\|_\beta \le \|x\|_\alpha^{\frac{\gamma-\beta}{\gamma-\alpha}} \|x\|_\gamma^{\frac{\beta-\alpha}{\gamma-\alpha}}, \quad \alpha < \beta < \gamma, \tag{6.15}$$

which follows immediately from the definition of the norms in the different spaces.

In §6.2 we will solve the stochastic equation (6.1) and in §6.3 we shall prove some useful estimates for the solution. §6.4 will be devoted to existence of an invariant measure $\nu$ and §6.5 to studying the Kolmogorov equation in $L^2(H, \nu)$.

## 6.2 Solution of the stochastic equation

Existence and uniqueness of the solution of (6.1) has been proved in [59] by using Galerkin approximations. We will follow here the proof given in [50]. For this we write equation (6.8) in the usual mild form,

$$X(t,x) = e^{tA}x + \int_0^t e^{(t-s)A}b(X(s,x))ds + W_A(t), \quad t \ge 0, \tag{6.16}$$

where $W_A(t)$ is the stochastic convolution

$$W_A(t) = \int_0^t e^{(t-s)A}dW(s), \quad t \ge 0.$$

We are going to solve equation (6.16) by a fixed point argument in the space $L^4(0, T; L^4_\#)$.

**Lemma 6.4.** *Define the mapping*

$$\Gamma(f) = \int_0^t e^{(t-s)A}b(f(s))ds, \quad f \in L^4(0, T; L^4_\#),\ t \ge 0.$$

*Then $\Gamma$ maps $L^4(0, T; L^4_\#)$ into itself and for any $f, g \in L^4(0, T; L^4_\#)$ we have*

$$|\Gamma(f) - \Gamma(g)|_{L^4(0,T;L^4_\#)} \le 4(|f|_{L^4(0,T;L^4_\#)} + |g|_{L^4(0,T;L^4_\#)})\, |f - g|_4, \quad x, y \in L^4_\#.$$

*Proof. Step* 1. The following inclusion holds:

$$L^\infty(0, T; H) \cap L^2(0, T; H^1_\#) \subset L^4(0, T; L^4_\#). \tag{6.17}$$

Let in fact $x \in L^\infty(0, T; H^1_\#)$. Then by the Sobolev embedding (6.14) it follows that

$$|x(t)|_4 \le c_1 \|x(t)\|_{1/2}, \quad t \in [0, T].$$

Now, using the interpolatory estimate (6.15) we find that

$$|x(t)|_4 \le c_1 |x(t)|_2^{1/2} \|x\|_1^{1/2}, \quad t \in [0, T].$$

Finally, taking the fourth power of both sides of this identity and integrating with respect to $t$ over $[0, T]$ yields

$$\int_0^T |x(t)|_4^4 dt \leq c_1^4 \int_0^T |x(t)|_2^2 \, \|x(t)\|_1^2 dt \leq c^4 |x|_{L^\infty(0,T;H)}^2 \, |x|_{L^2(0,T;H_\#^1)}^2.$$

Since $L^\infty(0, T; H_\#^1)$ is dense in $L^\infty(0, T; H) \cap L^2(0, T; H_\#^1)$, Step 1 follows.

*Step* 2. Conclusion.

Let $f \in L^4(0, T; L_\#^4)$. Then by Proposition 6.2 it follows that that $b(f)$ belong to $L^2(0, T; H_\#^{-1})$. Thus, by a well-known result on abstract evolution equations, see [77], we have that $\Gamma(f) \in L^\infty(0, T; H) \cap L^2(0, T; H_\#^1)$. Consequently, by Step 1, we have that $\Gamma(f) \in L^4(0, T; L_\#^4)$. Finally, (6.17) follows from (6.12).     □

We denote by $L_W^4([0, T]; L_\#^4)$ the space of all adapted processes $X$ in $(\Omega, \mathscr{F}, \mathbb{P})$ taking values in $L_\#^4$ such that

$$|X|_{L_W^4([0,T];L_\#^4)}^4 := \mathbb{E}\left(\int_0^T |X(s)|_4^4 ds\right) < +\infty.$$

Let us prove the main result of the present section, see [50].

**Theorem 6.5.** *For any $x \in H$ there exists a unique solution $X(\cdot, x) \in L_W^4([0, T]; L_\#^4)$ of the equation* (6.16).

*Proof.* Write equation (6.16) as

$$X(\cdot, x) = e^{\cdot A}x + \Gamma(X(\cdot, x)) + W_A(\cdot, x).$$

Notice that $e^{\cdot A}x \in L^4([0, T]; L_\#^4)$ by [77] and that $W_A(\cdot, x) \in L_W^4([0, T]; L_\#^4)$ by Theorem 2.13. Then the local existence follows from a classical argument, since $\Gamma$ is locally Lipschitz continuous by Lemma 6.4. So, to prove the global existence it is enough to find an a priori estimate for the solution $X(t, x)$ in $L^\infty(0, T; H) \cap L^2(0, T; H_\#^1)$ which is included in $L^4(0, T; L_\#^4)$ by (6.17). For this we set $y(t) = X(t, x) - W_A(t)$, so that (6.16) reduces to the deterministic problem

$$\begin{cases} \dfrac{dy(t)}{dt} = Ay(t) + b(y(t) + W_A(t)), & t \in [0, T], \\[2mm] y(0) = x. \end{cases} \tag{6.18}$$

We proceed now formally assuming that $Y(t)$ is a strict solution of (6.18) (we can reduce to this case by a suitable approximation). Multiplying the first equation in (6.18) by $y(t)$ and integrating over $\mathcal{O}$ we obtain, using Proposition 6.1,

$$\frac{1}{2}\frac{d}{dt}\,|y(t)|_2^2 + \|y(t)\|_1^2 \;=\; b(y(t) + W_A(t), y(t) + W_A(t), y(t))$$

$$=\; b(y(t) + W_A(t), W_A(t), y(t))$$

$$=\; -b(y(t), y(t), W_A(t)) - b(W_A(t), y(t), W_A(t)).$$

Let us first estimate $b(y(t), y(t), W_A(t))$. By (6.9) we have

$$|b(y(t), y(t), W_A(t))|_2 \leq 4\|y(t)\|_1\,|y(t)|_4\,|W_A(t)|_4.$$

Using the Sobolev embedding (6.14), we obtain

$$|b(y(t), y(t), W_A(t))| \leq 4c_1\|y(t)\|_1\,\|y(t)\|_{\frac{1}{2}}\,|W_A(t)|_4.$$

Moreover, by the interpolatory estimate (6.15) we find

$$|b(y(t), y(t), W_A(t))| \leq 4c_1|y(t)|_2^{1/2}\,\|y(t)\|_1^{3/2}\,|W_A(t)|_4.$$

Finally this implies that for a suitable constant $c_2$,

$$|b(y(t), y(t), W_A(t))|_2 \leq \frac{1}{4}\|y(t)\|_1^2 + c_2\,|y(t)|_2^2\,\|W_A(t)\|_4^4,\ t \in [0, T]. \qquad (6.19)$$

Let us now estimate $b(W_A(t), y(t), W_A(t))$. By (6.9) we have

$$|b(W_A(t), y(t), W_A(t))|_2 \leq 4\|y(t)\|_1\,|W_A(t)|_4^2,$$

and therefore

$$|b(W_A(t), y(t), W_A(t))| \leq \frac{1}{4}\|y(t)\|_1^2 + 16\,|W_A(t)|_4^4,\ t \in [0, T]. \qquad (6.20)$$

Adding the inequalities (6.19) and (6.20) we get

$$\frac{d}{dt}\,|y(t)|_2^2 + \|y(t)\|_1^2 \leq K_1|y(t)|_2^2\,|W_A(t)|_4^4 + K_2|W_A(t)|_4^4, \qquad (6.21)$$

where $K_1, K_2$ are suitable constants. By the usual comparison argument we find

$$|y(T)|_2^2 + \int_0^T e^{K_1 \int_s^T |W_A(r)|_4^4 dr}\,\|y(s)\|_1^2 ds$$

$$\leq e^{K_1 \int_0^T |W_A(r)|_4^4 dr}\,|x|_2^2 + K_2 \int_0^T e^{K_1 \int_s^T |W_A(r)|_4^4 dr}\,|W_A(s)|_4^4 ds.$$

$$(6.22)$$

This yields the required a-priori estimate. The proof is complete. $\qquad\qquad \square$

## 6.3  Estimates for the solution

In this section we gather for further use several estimates on the solution $X(t, x)$ of equation (6.8) and of its derivative with respect to $x$. The basic tool will be Itô's formula which for the sake of brevity we shall apply without any justification (however, it could be proved by suitable approximations as in previous chapters). We shall follow [9].

**Lemma 6.6.** *We have*

$$\mathbb{E}|X(t,x)|_2^2 + \mathbb{E}\int_0^t \|X(s,x)\|_1^2 ds = |x|_2^2 + t\,\mathrm{Tr}\,[(-A)^{-\gamma}], \quad t \geq 0. \qquad (6.23)$$

*Proof.* By the Itô formula we obtain

$$\mathbb{E}|X(t,x)|_2^2 = |x|^2 + \mathbb{E}\int_0^t \left[\langle AX(s,x) + b(X(s,x), X(s,x))_2\rangle\right] ds + t\,\mathrm{Tr}\,[(-A)^{-\gamma}].$$

Since $\langle AX(s,x), X(s,x)\rangle_2 = -\|X(s,x)\|_1^2$ and $\langle b(X(s,x), X(s,x))_2 = 0$ by Proposition 6.1, the conclusion follows.                                                                   □

**Lemma 6.7.** *Assume that $\alpha \leq \min\{\frac{1}{\|C\|}, 1\}$. Then we have*

$$\mathbb{E}\left(e^{\alpha\int_0^t |AX(s,x)|_2^2 ds}\right) \leq e^{\alpha\|x\|_1^2}\, e^{t\,\mathrm{Tr}\,[(-A)C]}, \quad x \in H,\ t \geq 0. \qquad (6.24)$$

*Proof.* Fix $\alpha \leq \frac{1}{\|C\|}$. We shall apply the Itô formula to the process $e^{\alpha V(t)}$ where

$$V(t) = \|X(t,x)\|_1^2 + \int_0^t |AX(s,x)|_2^2 ds.$$

First notice that

$$d\|X(t,x)\|_1^2 = -2\langle AX(t,x), AX(t,x) + b(X(t,x))\rangle_2 dt$$
$$+ \mathrm{Tr}\,[(-A)C] - \langle AX(t,x), \sqrt{C}dW(t)\rangle_2.$$

Therefore, taking into account (6.13), we obtain

$$d\|X(t,x)\|_1^2 = -2|AX(t)|_2^2 dt + \mathrm{Tr}\,[(-A)C] - \langle AX(t), \sqrt{C}dW(t)\rangle.$$

Consequently

$$dV(t) = -|AX(t,x)|_2^2 dt + \mathrm{Tr}\,[(-A)C]dt - \langle AX(t,x), \sqrt{C}dW(t)\rangle.$$

Moreover, again by Itô's formula, we have

$$de^{\alpha V(t)} = \alpha e^{\alpha V(t)}dV(t) + \frac{1}{2}\,\alpha^2 e^{\alpha V(t)}\,|\sqrt{C}AX(t,x)|_2^2 dt.$$

So,

$$
\begin{aligned}
de^{\alpha V(t)} &= \alpha e^{\alpha V(t)}(-|AX(t,x)|_2^2 dt + \text{Tr}\,[(-A)C]dt - \langle AX(t,x), \sqrt{C}dW(t)\rangle) \\
&\quad + \frac{1}{2}\alpha^2 e^{\alpha V(t)}\,|\sqrt{C}AX(t,x)|_2^2 dt \\
&\leq \frac{1}{2}\alpha e^{\alpha V(t)}\,(\alpha\|C\| - 2)|AX(t,x)|_2^2 dt \\
&\quad + \alpha e^{\alpha V(t)}\,(\text{Tr}\,[(-A)C]dt - \langle AX(t,x)), \sqrt{C}dW(t)\rangle) \\
&\leq \alpha\,\text{Tr}\,[(-A)C]e^{\alpha V(t)}dt - \langle AX(t), \sqrt{C}dW(t)\rangle)e^{\alpha V(t)},
\end{aligned}
$$

since $\alpha \leq \frac{1}{\|C\|}$. Consequently, we find that

$$
\mathbb{E}\left(e^{\alpha V(t)}\right) \leq e^{\alpha|x|_2^2} + \text{Tr}\,[(-A)C]\int_0^t \mathbb{E}\left(e^{\alpha V(s)}\right)ds
$$

and the conclusion follows from the Gronwall lemma. $\qquad\square$

We now consider some estimates for the derivative $\eta^h(t,x) = DX(t,x)\cdot h$ of $X(t,x)$ where $h \in H$. $\eta^h(t,x)$ is the mild solution of the problem

$$
\begin{cases}
\dfrac{d}{dt}\,\eta^h(t,x) = A\eta^h(t,x) + D_\xi X(t,x)\cdot\eta^h(t,x) + D_\xi\eta^h(t,x)\cdot X(t,x), \\[2mm]
\eta_\varepsilon^h(0,s) = h.
\end{cases}
\tag{6.25}
$$

**Lemma 6.8.** *There exists $\kappa > 0, q \in (1,2)$ such that*

$$
|\eta^h(t,x)|_2^2 \leq e^{\kappa\int_0^t |AX(s,x)|_2^q ds}\,|h|_2^2, \quad x,h \in H.
\tag{6.26}
$$

*Proof.* Let $q \in (1,2)$ and set $p = \frac{q}{q-1}$. Multiplying both sides of the first equation of (6.25) by $\eta^h(t,x)$ and taking into account (6.13) yields

$$
\begin{aligned}
\frac{1}{2}\frac{d}{dt}\,|\eta^h(t,x)|_2^2 + \|\eta^h(t,x)\|_1^2 &= b(\eta^h(t,x), D_\xi X(t,x), \eta^h(t,x)) \\
&\leq |D_\xi X(t,x)|_p\,|(\eta^h(t,x))^2|_q.
\end{aligned}
\tag{6.27}
$$

By the Sobolev embedding theorem there exists $c > 0$ such that

$$
|D_\xi X(t,x)|_p \leq c|AX(t,x)|_2
$$

and

$$
|\eta^h(t,x)|_{2q} \leq c\|\eta^h(t,x)\|_{1-\frac{1}{q}}.
$$

Moreover, in view of (6.15),

$$|\eta^h(t,x)|_{1-\frac{1}{q}} \le c|\eta^h(t,x)|_2^{\frac{1}{q}} \, \|\eta^h(t,x)\|_1^{1-\frac{1}{q}}$$

and consequently

$$|D_\xi X(t,x)|_p \, |(\eta^h(t,x))^2|_q \le c|AX(t,x)|_2 \, |\eta^h(t,x)|_2^{\frac{2}{q}} \, \|\eta^h(t,x)\|_1^{2(1-\frac{1}{q})}.$$

Since for any $a, b \ge 0$ the inequality

$$ab \le \frac{1}{q} a^q + \left(1 - \frac{1}{q}\right) b^{\frac{q}{q-1}}$$

holds, there exists $c' > 0$ such that

$$|D_\xi X(t,x)|_p \, |(\eta^h(t,x))^2|_q \le c'|AX(t,x)|_2^q \, |\eta^h(t,x)|_2^2 + \frac{1}{2} \, \|\eta^h(t,x)\|_2^2.$$

Then by (6.27) we obtain that

$$\frac{d}{dt} \, |\eta^h(t,x)|_2^2 \le 2c'|AX(t,x)|_2^q \, |\eta^h(t,x)|_2^2$$

and the conclusion follows.                                                                 □

The following corollary is a straightforward consequence of Lemma 6.8.

**Corollary 6.9.** *For any $\sigma > 0$ there exists $\kappa_\sigma > 0$ such that*

$$|\eta^h(t,x)|^2 \le e^{\kappa_\sigma t + \sigma \int_0^t |AX(s,x)|_2^2 ds} \, |h|_2^2, \quad x, h \in H, \ t \ge 0. \qquad (6.28)$$

*Proof.* It is enough to notice that for any $\sigma > 0$ there exists $\kappa_\sigma > 0$ such that

$$|AX(s,x)|_2^q \le \sigma|AX(s,x)|_2^2 + \kappa_\sigma, \quad x \in H, \ s \ge 0. \qquad \Box$$

Now by Corollary 6.9 and Lemma 6.7 we obtain immediately the following basic estimate.

**Corollary 6.10.** *For any $\alpha \le \frac{1}{\|C\|}$ there exists $\omega_\alpha > 0$ such that*

$$\mathbb{E}\left(|\eta^h(t,x)|_2^2\right) \le e^{2\omega_\alpha t} \, e^{\alpha\|x\|_1^2} \, |h|_2^2, \quad t \ge 0 \quad x, h \in H. \qquad (6.29)$$

## 6.4  Invariant measure $\nu$

Let us consider the transition semigroup

$$P_t\varphi(x) = \mathbb{E}[\varphi(X(t,x))], \quad \varphi \in C_b(H), \ x \in H, \ t \ge 0.$$

**Proposition 6.11.** *There exists an invariant measure for the semigroup $P_t$.*

*Proof.* By Lemma 6.6 it follows that

$$\frac{1}{t}\,\mathbb{E}\int_0^t \|X_\varepsilon(s,x)\|_1^2 ds \le \frac{1}{t}\,|x|^2 + \operatorname{Tr}\left[(-A)^{-\gamma}\right], \quad t \ge 0. \qquad (6.30)$$

Let $x$ be fixed and let $\nu_{t,x}$ be the law of $X(t,x)$ and set

$$\mu_T = \frac{1}{T}\int_0^T \nu_{t,x} dt.$$

Let $B_{R,1}$ be the unitary ball in $H_\#^1$. Since the embedding of $D(A)$ into $H$ is compact, we have that $B_{R,1}$ is a compact subset of $H$. For any $R > 0$ we have, taking into account 6.30,

$$\mu_T(B_{R,1}^c) = \frac{1}{T}\int_0^T \nu_{t,x}(B_{R,1}^c)dt \le \frac{1}{TR^2}\int_0^T \mathbb{E}(|X(t,x)|2)dt$$

$$\le \frac{1}{R^2}(|x|^2 \operatorname{Tr}\left[(-A)^{-\gamma}\right]), \quad T > 1.$$

Therefore we have the sequence of measures $\{\mu_T\}_{T\ge 1}$, so that, by the Krylov–Bogoliubov theorem, there exists an invariant measure $\nu$ for $P_t$. $\qquad\square$

### 6.4.1 Estimates of some integral

We want here, following [9], to estimate the integral

$$\int_H |Ax|_2^2\, \nu(dx),$$

where $\nu$ is an invariant measure for $P_t$.

We shall denote by $K$ the Kolmogorov operator

$$K\varphi(x) = L\varphi(x) + \langle b_\varepsilon(x), D\varphi(x)\rangle, \quad x \in H, \ \varphi \in \mathscr{E}_A(H),$$

where $L$ is the Ornstein–Uhlenbeck operator

$$L\varphi(x) = \frac{1}{2}\operatorname{Tr}\left[CD^2\varphi(x)\right] + \langle Ax, D\varphi(x)\rangle, \quad x \in H, \ \varphi \in \mathscr{E}_A(H).$$

**Proposition 6.12.** *Let $\delta < \delta_0 := \frac{1}{\|C\|}$. Then, there exists $c > 0$ independent of $\delta$ such that*

$$\int_H |Ax|_2^2\, e^{\delta\|x\|_1^2}\nu(dx) \le c. \qquad (6.31)$$

*Proof.* Let us compute $K\varphi$ for $\varphi(x) = e^{\delta\|x\|_1^2}$ with $\delta < \delta_0$. (Notice that $\varphi$ does not belong to $\mathscr{E}_A(H)$, however it can be easily approximated by functions of $\mathscr{E}_A(H)$.)

We have

$$D\varphi(x) = -2\delta e^{\delta\|x\|_1^2} Ax, \quad D^2\varphi(x) = 4\delta^2 e^{\delta\|x\|_1^2} Ax \otimes Ax - 2\delta e^{\delta\|x\|_1^2} A.$$

It follows that

$$\begin{aligned} K\varphi &= (\delta \operatorname{Tr}\left[(-A)C\right] + 2\delta^2|C^{1/2}Ax|_2^2 - 2\delta|Ax|_2^2)e^{\delta\|x\|_1^2} \\ &\leq (\delta \operatorname{Tr}\left[(-A)C\right] + 2\delta^2\|C\|\,|Ax|_2^2 - 2\delta|Ax|_2^2)e^{\delta\|x\|_2^2}. \end{aligned}$$

Since $\int_H K\varphi d\nu = 0$, due to the invariance of $\nu$, we have

$$\int_H |Ax|_2^2\, e^{\delta\|x\|_1^2}\nu(dx) \leq \frac{\operatorname{Tr}\left[(-A)C\right]}{2(1 - \delta\|C\|)} \int_H e^{\delta\|x\|_1^2}\nu(dx). \tag{6.32}$$

It follows that for any $R > 0$ we have, using the Chebyshev inequality,

$$\begin{aligned} \int_H e^{\delta\|x\|_1^2}\nu(dx) &= \int_{\{\|x\|_1 \leq R\}} e^{\delta\|x\|_1^2}\nu(dx) + \int_{\{\|x\|_1 > R\}} e^{\delta\|x\|_1^2}\nu(dx) \\ &\leq e^{\delta R^2} + \frac{1}{R^2} \int_H \|x\|_1^2 e^{\delta\|x\|_1^2}\nu(dx) \\ &\leq e^{\delta R^2} + \frac{1}{R^2} \int_H |Ax|_2^2 e^{\delta\|x\|_1^2}\nu(dx). \end{aligned}$$

Now, taking into account (6.32), it follows that

$$\int_H e^{\delta\|x\|_1^2}\nu(dx) \leq e^{\delta R^2} + \frac{\operatorname{Tr}\left[(-A)C\right]}{2R^2(1 - \delta\|C\|)} \int_H e^{\delta\|x\|_1^2}\nu(dx)$$

and choosing

$$R^2 = \frac{\operatorname{Tr}\left[(-A)C\right]}{2(1 - \delta\|C\|)},$$

we see that

$$\int_H e^{\delta\|x\|_1^2}\nu(dx) \leq 2e^{\frac{\delta \operatorname{Tr}\left[(-A)C\right]}{2(1 - \delta\|C\|)}}.$$

Finally, the conclusion follows using again (6.32).  $\square$

## 6.5  Kolmogorov equation

We are still concerned with the transition semigroup $P_t$ and with a fixed invariant measure $\nu$ of $P_t$. $P_t$ can be uniquely extended to a strongly continuous semigroup of contractions in $L^2(H, \nu)$, still denoted by $P_t$. We shall denote by $K_2$ its infinitesimal generator. As in the previous sections, we are going to prove that $K_2$ is the closure of the Kolmogorov operator

$$K_0(x) = L\varphi(x) + \langle b(x), D\varphi(x) \rangle, \quad x \in H, \ \varphi \in \mathscr{E}_A(H),$$

where $L$ is the Ornstein–Uhlenbeck generator.

It is convenient to introduce the approximating problems

$$\begin{cases} dX_\varepsilon(t) = (AX_\varepsilon(t) + b_\varepsilon(X_\varepsilon(t)))dt + \sqrt{C}dW(t), \\ X_\varepsilon(0) = x \in H, \end{cases} \tag{6.33}$$

where

$$b_\varepsilon(x) = \begin{cases} b(x) & \text{if } \|x\|_1 \leq \dfrac{1}{\varepsilon}, \\[2mm] \dfrac{b(x)}{\varepsilon^2 \|x\|_1^2} & \text{if } \|x\|_1 > \dfrac{1}{\varepsilon}. \end{cases} \tag{6.34}$$

Since, for $\varepsilon > 0$, $b_\varepsilon$ is regular and bounded in $H^1_\#$, one can show, proceeding as for the proof of Theorem 6.5 , that problem (6.33) has a unique solution which we denote by $X_\varepsilon(t, x)$. We shall denote by $K_\varepsilon$ the Kolmogorov operator

$$K_\varepsilon(x) = L\varphi(x) + \langle b_\varepsilon(x), D\varphi(x) \rangle, \quad x \in H, \ \varphi \in \mathscr{E}_A(H).$$

We first prove that $K_2$ is an extension of $K_0$.

**Proposition 6.13.** *For any $\varphi \in \mathscr{E}_A(H)$, we have $\varphi \in D(K_2)$ and $K_2\varphi = K_0\varphi$.*

*Proof.* Let $\varphi \in \mathscr{E}_A(H)$. Then by the Itô formula we have that

$$\lim_{h \to 0} \frac{1}{h} (P_h\varphi(x) - \varphi(x)) = N_0\varphi(x), \quad x \in H.$$

To prove that $\varphi \in D(K_2)$ it is enough to show that

$$\frac{1}{h} (P_h\varphi(x) - \varphi(x))$$

is equibounded in $L^2(H, \nu)$ for $h \in (0, 1]$. We have in fact for any $x \in H$,

$$|P_h\varphi(x) - \varphi(x)| \leq \|\varphi\|_1 \int_0^h \mathbb{E}(|AX(s, x)|_2 + |b(X(s, x))|_2)ds.$$

By the Hölder inequality we find that

$$|P_h\varphi(x) - \varphi(x)|^2 \;\leq\; \|\varphi\|_1^2 \, h \int_0^h [\mathbb{E}(|AX(s,x)|_2 + |b(X(s,x))|_2)]^2 \, ds$$

$$\leq\; 2\|\varphi\|_1^2 \, h \int_0^h \left[ \mathbb{E}(|AX(s,x)|_2^2) + \mathbb{E}(|b(X(s,x))|_2^2) \right] ds$$

$$=\; 2\|\varphi\|_1^2 \, h \int_0^h \left[ P_s(|A \cdot |_2^2)(x) + P_s(|b(\cdot)|_2^2)(x) \right] ds.$$

Integrating with respect to $\nu$ over $H$ and taking into account the invariance of $\nu$, it follows that

$$\|P_h\varphi - \varphi\|_{L^2(H,\nu)}^2 \leq 2h^2 \|\varphi\|_1^2 \int_H (|Ax|_2^2 + |b(x)|_2^2)\nu(dx) < +\infty,$$

thanks to (6.31).                                                                 □

Finally, we can prove the following result.

**Theorem 6.14.** *Let $\nu$ be an invariant measure for $P_t$ and $K_2$ the infinitesimal generator of $P_t$ in $L^2(H,\nu)$. Then $K_2$ is the closure of $K_0$ in $L^2(H,\nu)$. Moreover, if $\varphi \in D(K_2)$, then $|C^{1/2}D\varphi|_2 \in L^2(H,\nu)$ and*

$$\int_H K_2\varphi \, \varphi d\nu = -\frac{1}{2} \int_H |C^{1/2}D\varphi|_2^2 d\nu. \tag{6.35}$$

*Proof.* We have to prove that the closure $\overline{K_0}$ of $K_0$ coincides with $K_2$. Let $f \in C_b^1(H)$ and $\lambda > 0$. Since $b_\varepsilon$ is bounded and regular, there exists a unique solution $\varphi_\varepsilon \in D(L) \cap C_b^1(H)$ of the equation

$$\lambda\varphi_\varepsilon(x) - L\varphi_\varepsilon(x) - \langle b_\varepsilon(x), D\varphi_\varepsilon(x)\rangle = f(x), \quad x \in H,$$

given by

$$\varphi_\varepsilon(x) = \int_0^\infty e^{-\lambda t} \mathbb{E}[f(X_\varepsilon(t,x))]dt, \quad x \in H.$$

Fix $\alpha \in (0, \|C\|^{-1})$. Then, in view of Corollary 6.10, it follows that for any $\lambda > \omega_\alpha$ we have

$$\langle D\varphi_\varepsilon(x), h\rangle = \mathbb{E} \int_0^\infty e^{-\lambda t} \langle Df(X_\varepsilon(t,x)), \eta_\varepsilon^h(t,x)\rangle dt, \quad x, h \in H$$

and

$$|\langle D\varphi_\varepsilon(x), h\rangle| \leq \frac{1}{\lambda - \omega_\alpha} e^{\frac{\alpha}{2}\|x\|^2} \|Df\|_0 \, |h|^2, \quad h, x \in H.$$

Thus, by the arbitrariness of $h$, it follows that

$$|D\varphi_\varepsilon(x)| \leq \frac{1}{\lambda - \omega_\alpha} e^{\frac{\alpha}{2}\|x\|^2} \|Df\|_0, \quad x \in H. \tag{6.36}$$

We now fix $\lambda > \omega_\alpha$.

*Claim 1.* We have $\varphi_\varepsilon \in D(\overline{K_0})$ and

$$\lambda\varphi_\varepsilon - \overline{K_0}\varphi_\varepsilon = \langle b_\varepsilon(x) - b(x), D\varphi_\varepsilon \rangle + f.$$

In fact, by Corollary 2.71, there exists a three-index sequence $\{\psi_{n_1,n_2,n_3}\} \subset \mathscr{E}_A(H)$ such that

$$\lim_{n_1\to\infty} \lim_{n_2\to\infty} \lim_{n_3\to\infty} \psi_{n_1,n_2,n_3}(x) = \varphi_\varepsilon(x), \quad x \in H,$$

$$\lim_{n_1\to\infty} \lim_{n_2\to\infty} \lim_{n_3\to\infty} D\psi_{n_1,n_2,n_3}(x) = D\varphi_\varepsilon(x), \quad x \in H,$$

$$\lim_{n_1\to\infty} \lim_{n_2\to\infty} \lim_{n_3\to\infty} L\psi_{n_1,n_2,n_3}(x) = L\varphi_\varepsilon(x), \quad x \in H.$$

By Proposition 6.12 and the dominated convergence theorem, it follows that

$$\lim_{n_1\to\infty} \lim_{n_2\to\infty} K_0\psi_{n_1,n_2,n_3}(x) = \overline{K_0}\varphi_\varepsilon(x) = L\varphi_\varepsilon + \langle B_\varepsilon(x), D\varphi_\varepsilon \rangle, \quad x \in H.$$

So, Claim 1 is proved.

*Claim 2.* We have

$$\lim_{\varepsilon\to 0} \langle b_\varepsilon(x) - b(x), D\varphi_\varepsilon \rangle = 0 \quad \text{in } L^2(H, \nu).$$

Once Claim 2 is proved, we deduce that the closure of the range of $\lambda - \overline{K_0}$ is dense in $L^2(H, \nu)$ and so, in view of the Lumer–Phillips Theorem 3.20, that $\overline{K_0} = K$ as claimed.

To prove Claim 2, we notice that, taking into account (6.36),

$$\int_H |\langle b_\varepsilon(x) - b(x), D\varphi_\varepsilon \rangle|^2 d\nu = \int_{\{\|x\|_1 \geq 1/\varepsilon\}} |\langle b_\varepsilon(x) - b(x), D\varphi_\varepsilon \rangle|^2 d\nu$$

$$\leq \frac{1}{(\lambda - \omega_\alpha)^2} \|Df\|_0^2 \int_{\{\|x\|_1 \geq 1/\varepsilon\}} |b(x)|_2^2 \frac{\varepsilon\|x\|_1^2 - 1}{\varepsilon\|x\|_1^2} e^{\alpha|x|_1^2} d\nu$$

$$\leq \frac{1}{(\lambda - \omega_\alpha)^2} \|Df\|_0^2 \int_{\{\|x\|_1 \geq 1/\varepsilon\}} |b(x)|_2^2 e^{\alpha|x|_2^2} d\nu.$$

Thus, it is enough to show that

$$\int_H |b(x)|_2^2 e^{\alpha|x|_2^2} \nu(dx) < +\infty. \tag{6.37}$$

We have in fact

$$|b(x)|_2^2 \leq \int_{\mathcal{O}} |x|_2^2 \, |x_\xi|_2^2 d\xi \leq |Ax|_2^2 \, \|x\|_1^2.$$

It follows that if $\alpha' < \alpha$,

$$\int_H |\langle B(x), D\varphi\rangle|^2 \nu(dx) \leq \int_H |Ax|_2^2 \, \|x\|_1^2 e^{\alpha'|x|_2^2} d\nu \leq c \int_H |Ax|_2^2 \, e^{\alpha|x|_2^2} d\nu,$$

and the conclusion follows from (6.33). Finally, (6.35) follows integrating with respect to $\nu$ the identity

$$K_0(\varphi^2) = 2\varphi K_0\varphi + |Q^{1/2}D\varphi|^2.$$

The proof is complete.                                        □

# Bibliography

[1] R. A. Adams, *Sobolev Spaces*, Academic Press, 1975.

[2] S. A. Agmon, *Lectures on Elliptic Boundary Value Problems*, Van Nostrand, 1965.

[3] S. Albeverio and A. B. Cruzeiro, *Global flows with invariant (Gibbs) measures for Euler and Navier–Stokes two dimensional fluids*, Commun. Math. Phys. **129**, 431–444, 1990.

[4] S. Albeverio and M. Röckner, *New developments in the theory and applications of Dirichlet forms in Stochastic processes, Physics and Geometry*, S. Albeverio et al. eds, World Scientific, Singapore, 27–76, 1990.

[5] S. Albeverio and M. Röckner, *Stochastic differential equations in infinite dimensions: solutions via Dirichlet forms*, Probab. Theory Relat. Fields, 89, 347–386, 1991.

[6] V. Barbu and G. Da Prato, *The stochastic nonlinear damped wave equation*, Appl. Math. Optimiz. 46, 125–141, 2002.

[7] V. Barbu and G. Da Prato, *The two phase stochastic Stefan problem*, Probab. Theory Relat. Fields, **124**, 544–560, 2002.

[8] V. Barbu, G. Da Prato and A. Debussche, *The transition semigroup of stochastic Navier–Stokes equations in 2-D*, Infinite Dimensional Analysis, Quantum Probability and related topics, vol. 7. n.2, 163–181, 2004.

[9] V. Barbu, G. Da Prato and A. Debussche, *m-dissipativity of Kolmogorov operators corresponding to periodic 2D Navier Stokes equations*, Rend. Mat. Acc. Lincei, s.9. v.15: 29–38, 2004.

[10] J. M. Bismut, *Large deviations and the Malliavin calculus*, Birkhäuser, 1984.

[11] V. Bogachev, G. Da Prato and M. Röckner, *Regularity of invariant measures for a class of perturbed Ornstein–Uhlenbeck operators*, Nonlinear Diff. Equations Appl., **3**, 261–268, 1996.

[12] V.I. Bogachev, N.V. Krylov and M. Röckner, *On regularity of transition probabilities and invariant measures of singular diffusions under minimal conditions*, Comm. Partial Diff. Equations, 26, no. 11–12, 2001.

[13] V. Bogachev and M. Röckner, *Regularity of invariant measures on finite and infinite dimensional spaces and applications*, J. Funct. Anal. 133, 168–223, 1995.

[14] V. Bogachev and M. Röckner, *Elliptic equations for measures on infinite dimensional spaces and applications*, Probab. Theory Relat. Fields, 120, 445–496, 2001.

[15] V. Bogachev, M. Röckner and B. Schmuland, *Generalized Mehler semigroups and applications*, Probab. Theory Relat. Fields **105**, 2, 193–225, 1996.

[16] H. Brézis, *Operatéurs maximaux monotones*, North-Holland, 1973.

[17] S. Cerrai, *A Hille–Yosida theorem for weakly continuous semigroups*, Semigroup Forum, **49**, 349–367, 1994.

[18] S. Cerrai, *Weakly continuous semigroups in the space of functions with polynomial growth*, Dyn. Syst. Appl., **4**, 351–372, 1995.

[19] S. Cerrai, *Second order PDE's in finite and infinite dimensions. A probabilistic approach*, Lecture Notes in Mathematics, **1762**, Springer-Verlag, 2001.

[20] S. Cerrai and M. Röckner, *Large deviations for stochastic reaction-diffusion systems with multiplicative noise and non-Lipschitz reaction term*, Annals of Probability 32, 1–40, 2004.

[21] A. Chojnowska-Michalik and B. Goldys, *Existence, uniqueness and invariant measures for stochastic semilinear equations*, Probab. Th. Relat. Fields, **102**, 331–356, 1995.

[22] A. Chojnowska-Michalik and B. Goldys, *On regularity properties of nonsymmetric Ornstein–Uhlenbeck semigroup in $L^p$ spaces*, Stoch. Stoch. Reports, **59**, 183–209, 1996.

[23] A. Chojnowska-Michalik and B. Goldys, *Nonsymmetric Ornstein–Uhlenbeck semigroup as a second quantized operator*, J. Math. Kyoto Univ., **36**, 481–498, 1996.

[24] A. Chojnowska-Michalik and B. Goldys, *Generalized Ornstein–Uhlenbeck semigroups: Littlewood–Paley–Stein inequalities and the P. A. Meyer equivalence of norms*, J. Functional Anal., **182**, 243–279, 2001.

[25] A. Chojnowska-Michalik and B. Goldys, *Symmetric Ornstein–Uhlenbeck semigroups and their generators*, Probab. Th. Relat. Fields, **124**, 459–486, 2002.

[26] H. Crauel, A. Debussche and F. Flandoli, *Random attractors*, J. Dynam. Differential Equations 9, 307–341, 1997.

[27] R. Dalang and N. Frangos, *The stochastic wave equation in two spatial dimensions*, Ann. Probab., **26**, 1, 187–212, 1998.

[28] Yu. Daleckij and S. V. Fomin, *Measures and differential equations in infinite-dimensional space*, Kluwer, 1991.

[29] G. Da Prato, *Null controllability and strong Feller property of Markov transition semigroups*, Nonlinear Analysis TMA, **25**, 9–10, 941–949, 1995.

[30] G. Da Prato, *Stochastic evolution equations by semigroups methods*, Centre de Recerca Matemàtica, Quaderns nùm 11/ gener 1998.

[31] G. Da Prato and A. Debussche, *Stochastic Cahn–Hilliard equation*, Nonlinear Anal. **26**, 2, 241–263, 1996.

[32] G. Da Prato and A. Debussche, *Maximal dissipativity of the Dirichlet operator corresponding to the Burgers equation*, CMS Conference Proceedings, vol. 28, Canadian Mathematical Society, 145–170, 2000.

[33] G. Da Prato and A. Debussche, *2D Navier–Stokes equations driven by a space-time white noise*, J. Functional Analysis, **196**, 1, 180–210, 2002.

[34] G. Da Prato and A. Debussche, *Ergodicity for the 3D stochastic Navier-Stokes equations*, Journal Math. Pures Appl. **82**, 877–947, 2003.

[35] G. Da Prato and A. Debussche, *Absolute continuity of the invariant measures for some stochastic PDEs*, Journal of Statistical Physics, **115**, no. 112, 451–468, 2004.

[36] G. Da Prato and A. Debussche, *m-dissipativity of Kolmogorov operators corresponding to Burgers equations with space-time white noise*, preprint SNS, 2004.

[37] G. Da Prato, A. Debussche and B. Goldys, *Invariant measures of non symmetric dissipative stochastic systems*, Probab. Theory Relat. Fields, **123**, 3, 355–380.

[38] G. Da Prato, A. Debussche and R. Temam, *Stochastic Burgers equation*, Nonlinear Differential Equations Appl., 389–402, 1994.

[39] G. Da Prato, A. Debussche and L. Tubaro, *Irregular semi-convex gradient systems perturbed by noise and application to the stochastic Cahn–Hilliard equation*, Ann. Institut Poincaré, 40, 1, 73–88, 2004.

[40] G. Da Prato, A. Debussche and L. Tubaro, *Coupling for some partial differential equations driven by white noise*, Preprint, 2004.

[41] G. Da Prato, D. Elworthy and J. Zabczyk, *Strong Feller property for stochastic semilinear equations,* Stoch. Anal. Appl., **13**, 35–45, 1995.

[42] G. Da Prato, S. Kwapien and J. Zabczyk, *Regularity of solutions of linear stochastic equations in Hilbert Spaces,* Stochastics, **23**, 1–23, 1987.

[43] G. Da Prato and M. Röckner, *Singular dissipative stochastic equations in Hilbert spaces,* Probab. Theory Relat. Fields, **124**, 2, 261–303, 2002.

[44] G. Da Prato and M. Röckner, *Invariant measures for a stochastic porous medium equation,* in: "Stochastic Analysis and Related Topics in Kyoto" (H. Kunita, S. Watanabe and Y. Takahashi, eds.), Advanced Studies in Pure Mathematics, 41, AMS/Japan Math Society, 13–29, 2004.

[45] G. Da Prato and M. Röckner, *Weak solutions to stochastic porous media equations,* J. Evol. Equ, no. 2, 249–271, 2004.

[46] G. Da Prato and L. Tubaro, *A new method to prove self-adjointness of some infinite dimensional Dirichlet operator,* Probab. Theory Relat. Fields, **118**, 1, 131–145, 2000.

[47] G. Da Prato and L. Tubaro, *On a class of gradient systems with irregular potentials,* Infinite Dimensional Analysis, Quantum Probability and related topics, Vol. 4, 183–194, 2001.

[48] G. Da Prato and L. Tubaro, *Some results about dissipativity of Kolmogorov operators,* Czechoslovak Mathematical Journal, **51**, 126, 685–699, 2001.

[49] G. Da Prato and J. Zabczyk, *Stochastic equations in infinite dimensions,* Cambridge University Press, 1992.

[50] G. Da Prato and J. Zabczyk, *Ergodicity for infinite dimensional systems,* London Mathematical Society Lecture Notes, **229**, Cambridge University Press, 1996.

[51] G. Da Prato and J. Zabczyk, *Second Order Partial Differential Equations in Hilbert Spaces,* London Mathematical Society, Lecture Notes, **293**, Cambridge University Press, 2002.

[52] E.B. Davies, *One parameter semigroups,* Academic Press, 1980.

[53] J. D. Deuschel and D. Stroock, *Large deviations,* Academic Press, 1984.

[54] W. E., K. Khanin, A. Mazel and Y.G. Sinai, *Invariant measures for Burgers equation with stochastic forcing,* Ann. of Math. (2) 151, no. 3, 877–960, 2000.

[55] G. Lumer and R. S. Phillips, Dissipative operators in a Banach space, *Pac. J. Math.*, **11**, 679–698, 1961.

[56] W. E, J.C. Mattingly, Y. G. Sinai, *Gibbsian dynamics and ergodicity for the stochastically forced Navier–Stokes equation*, Commun. Math. Phys. **224**, 83–106, 2001.

[57] A. Eberle, *Uniqueness and non-uniqueness of singular diffusion operators*, Lecture Notes in Mathematics, **1718**, Springer-Verlag, 1999.

[58] K. D. Elworthy, *Stochastic flows on Riemannian manifolds*, Diffusion processes and related problems in analysis, Vol. II, M. A. Pinsky and V. Wihstutz (editors), 33–72, Birkhäuser, 1992.

[59] F. Flandoli, *Dissipativity and invariant measures for stochastic Navier–Stokes equations*, Nonlinear Differential Equations Appl., **1**, 403–423, 1994.

[60] F. Flandoli and F. Gozzi, *Kolmogorov equation associated to a stochastic Navier–Stokes equation*, J. Functional Anal., **160**, 312–336, 1998.

[61] F. Flandoli and B. Maslowski, *Ergodicity of the 2D Navier–Stokes equation under random perturbations*, Commun. Math. Phys. **171**, 119–141, 1995.

[62] M. Fuhrman, *Analyticity of transition semigroups and closability of bilinear forms in Hilbert spaces*, Studia Mathematica, **115**, 53–71, 1995.

[63] M. Fuhrman, *Hypercontractivité des semigroupes de Ornstein–Uhlenbeck non symétriques*, C. R. Acad. Sci. Paris, **321**, 929–932, 1995.

[64] M. Fukushima, Y. Oshima and M. Takeda, *Dirichlet forms and symmetric Markov processes*, de Gruyter, Berlin, 1994.

[65] D. Gątarek and B. Goldys. *On weak solutions of stochastic equations in Hilbert spaces*, Stochastics Stochastic Rep. **46**, 41–51, 1994.

[66] L. Gross, *Potential theory on Hilbert spaces*, J. Functional Anal., **1**, 123–181, 1967.

[67] L. Gross, *Logarithmic Sobolev inequalities*, Amer. J. Math., **97**, 1061–1083, 1976.

[68] M. Hairer, *Exponential mixing properties of stochastic PDEs through asymptotic coupling*, Probab. Theory Relat. Fields, **124**, 3, 345–380, 2000.

[69] R. S. Halmos, *Measure Theory*, Van Nostrand, 1950.

[70] I. Karatzas and E. Shreve, *Brownian motion and Stochastic Calculus*, Springer, 1991.

[71] R. Z. Khas'minskii, *Stochastic Stability of Differential Equations*, Sijthoff and Noordhoff, 1980.

[72] A. N. Kolmogorov, *Über die analytischen Methoden in der Wahrschein-lichkeitsrechnung*, Math. Ann., **104**, 415–458, 1931.

[73] S. Kuksin, A. Piatnitski and A. Shirikyan, *A coupling approach to randomly forced nonlinear PDEs. II*, Comm. Math. Phys. **230**, 1, 81–85, 2002.

[74] S. Kuksin and A. Shirikyan, *A coupling approach to randomly forced nonlinear PDEs. I*, Comm. Math. Phys. **221**, 1, 351–366, 2002.

[75] J. M. Lasry and P. L. Lions, *A remark on regularization in Hilbert spaces*, Israel J. Math., **55**, 257–266, 1986.

[76] P. Lévy, *Problèmes concrets d'analyse fonctionnelle*, Gauthier–Villar, 1951.

[77] J. L. Lions and E. Magenes, *Problèmes aux limites non homogènes et appli-cations*, Dunod., 1968

[78] V. Liskevich and M. Röckner, *Strong uniqueness for a class of infinite di-mensional Dirichlet operators and application to stochastic quantization*, Ann. Scuola Norm. Sup. Pisa Cl. Sci. (4),Vol. XXVII, pp. 69–91, 1998.

[79] G. Lumer and R. S. Phillips, *Dissipative operators in a Banach space*, Pac. J. Math., **11**, 679–698, 1961.

[80] A. Lunardi, *Analytic semigroups and optimal regularity in parabolic problems*, Birkhäuser, 1995.

[81] Z. M. Ma and M. Röckner, *Introduction to the Theory of (Non Symmetric) Dirichlet Forms*, Springer-Verlag, 1992.

[82] P. Malliavin, *Stochastic analysis*, Springer-Verlag, 1997.

[83] J.C. Mattingly, *Ergodicty of 2D Navier–Stokes equations with random forcing and large viscosity*, Commun. Math. Phys. **206**, 273–288, 1999.

[84] J.C. Mattingly, *Exponential convergence for the stochastically forced Navier-Stokes equations and other partially dissipative dynamics*, Comm. Math. Phys. **230**, no. 3, 421–462, 2002.

[85] J. Neveu, *Bases mathématiques du Calcul des Probabilités*, Masson, 1970.

[86] R. Mikulevicius and B. Rozovskii, *Stochastic Navier–Stokes equations for tur-bulent flows*, Preprint, University of Southern California, CAMS 01–09, 2001.

[87] A. Millet and P. L. Morien, *On a nonlinear stochastic wave equation in the plane: existence and uniqueness of the solution*, Ann. Appl. Probab. **11**, 922–951, 2001.

[88] A. Millet and M. Sanz-Solé, *Approximation and support theorem for a wave equation in two space dimensions*, Bernoulli **6**, 887–915, 2000.

[89] A. S. Nemirovski and S. M. Semenov, *The polynomial approximation of functions in Hilbert spaces*, Mat. Sb. (N.S.), **92**, 257–281, 1973.

[90] K. P. Parthasarathy, *Probability measures in metric spaces*, Academic Press, New York, 1967.

[91] K. Petersen, *Ergodic theory*, Cambridge Studies in Avanced Mathematics, 1983

[92] S. Peszat and J. Zabczyk, *Strong Feller property and irreducibility for diffusions on Hilbert spaces*, Ann. Probab., **23**, 157–172, 1995.

[93] A. Pietsch, *Nuclear locally convex spaces*, Springer-Verlag, 1972.

[94] E. Priola, *On a class of Markov type semigroups in spaces of uniformly continuous and bounded functions*, Studia Math., **136**, 271–295, 1999.

[95] M. Röckner, $L^p$-analysis of finite and infinite dimensional diffusions, Lecture Notes in Mathematics, **1715**, G. Da Prato (editor), Springer-Verlag, 65–116, 1999.

[96] W. Stannat, *The theory of generalized Dirichlet forms and its applications in Analysis and Stochastics*, Memoirs AMS, 678, 1999.

[97] D. W. Stroock and S.R.S Varhadan, *Multidimensional Diffusion Processes*, Springer-Verlag, 1979.

[98] J. Zabczyk, *Linear stochastic systems in Hilbert spaces: spectral properties and limit behavior*, Report **236**, Institute of Mathematics, Polish Academy of Sciences, 1981. Also in Banach Center Publications, **41**, 591–609, 1985.

[99] L. Zambotti, *A new approach to existence and uniqueness for martingale problems in infinite dimensions*, Probab. Th. Relat. Fields, **118**, 147–168, 2000.

# Index